Suffering and Evil in Nature

Suffering and Evil in Nature

Comparative Responses from Ecstatic Naturalism and Healing Cultures

Edited by
Joseph E. Harroff and Jea Sophia Oh

Foreword by Robert S. Corrington

LEXINGTON BOOKS
Lanham • Boulder • New York • London

Published by Lexington Books
An imprint of The Rowman & Littlefield Publishing Group, Inc.
4501 Forbes Boulevard, Suite 200, Lanham, Maryland 20706
www.rowman.com

86-90 Paul Street, London EC2A 4NE

British Library Cataloguing in Publication Information Available

Library of Congress Cataloging-in-Publication Data

Names: Harroff, Joseph E., editor. | Oh, Jea Sophia, editor.
Title: Suffering and evil in nature : comparative responses from ecstatic naturalism and healing cultures / edited by Joseph E. Harroff and Jea Sophia Oh ; foreword by Robert S. Corrington.
Description: Lanham, Maryland : Lexington Books, [2021] | Includes bibliographical references and index.
Identifiers: LCCN 2021038957 (print) | LCCN 2021038958 (ebook) | ISBN 9781793621740 (cloth) | ISBN 9781793621757 (ebook)
Subjects: LCSH: Nature—Religious aspects. | Philosophy of nature. | Suffering. | Good and evil. | Healing—Religious aspects.
Classification: LCC BL65.N35 S925 2021 (print) | LCC BL65.N35 (ebook) | DDC 214—dc23
LC record available at https://lccn.loc.gov/2021038957
LC ebook record available at https://lccn.loc.gov/2021038958

∞™ The paper used in this publication meets the minimum requirements of American National Standard for Information Sciences—Permanence of Paper for Printed Library Materials, ANSI/NISO Z39.48-1992.

Contents

Foreword

Query on the Edges of Ecstatic Naturalism

Robert S. Corrington

The ongoing practice of query has been called a process involving "inventive wonder." What does this mean? If inquiry seeks to answer questions in a specific domain, query has a more general drive to probe into the heart of nature and consciousness, thus examining numerous domains as they continue to emerge and fade away. Query does not provide a list of specific answers but seeks to find the questionable on the edges of thought and experience. The questionable per se lies beyond and beneath a cluster of detailed answers to finite modes of inquiry. Turning the question of query on its axis: What is inventive wonder? To be inventive is to allow possibilities and actualities to play within the encompassing of thought. Is wonder the same as mystery? Often a commitment to the concept of mystery can shut down query and provide an escape hatch for conceptual inertia, whereas a sense of wonder can quicken query and enhance creative probing into the nature of nature.

What are some of the nodal points for query on the edges of the perspective of ecstatic naturalism? One of the most challenging areas of query is to find a clear link between the Darwinian emergentism of finite consciousness and the Hindu notion of absolute consciousness that can be found at the point of origin. But what are the parameters of the concept of origin? Can origin be both ground and abyss? What type of ground and what type of abyss can be found? And if the Darwinian paradigm makes clear sense, how can that be woven together with a notion in which consciousness is of cosmic scope?

Further, how do evolution and involution interact? If evolution is the emergence and extinction of species, then involution lies at the heart of the travails of self-consciousness. What then is the link between involution and the possible return of cosmic consciousness? Avoiding teleological language from the side of evolution, what kind of teleology pertains to involution?

Does cosmic consciousness lose itself in nature, only to return in human self-consciousness and find itself anew?

Does a reframing of the concept of emanation bring clarity to an enriched understanding of involution and cosmic consciousness? Must there be only one line of emanation or can there be multiple emanations, perhaps unfolding in concert with each other? Asking a truncated spatial question: Where does emanation come from, assuming a nonliteral understanding of "come from"? The question of origin resurfaces in a new guise.

How helpful is a generalized understanding of the unconscious to illuminating the concepts of origin, cosmic consciousness, involution, and emanation? How many layers are there to the unconscious? One model posits four dimensions: personal, cultural, collective, and that of nature. Is the fourth dimension, that of the unconscious of nature, related to the dialectic of finite and cosmic consciousness? Does one of the keys lie in the drama of involution? Does involution dance on the edge of the finite to cosmic consciousness and back again?

In what ways is the human process (selving) a quickened mode of emanation? How does selving, as a fitful process of individuation, recapitulate origins and perhaps cosmic consciousness? If the selving process participates in all four dimensions of the unconscious, then how does entering into the unconscious aid in the expansion of consciousness, especially if this trajectory carries no guarantees?

The medieval (and later) distinction between *natura naturans* and *natura naturata* is used to explain the origin to a consequent—natural fecundity to the emergent realms of nature. The latter dimension is more obvious, but how does query enter into an articulation of *natura naturans*? If *natura naturans* can be seen as the unconscious of nature, how much of the unconscious can be known? Is the unconscious sheerly other to consciousness, its abjection, or can this otherness be overcome in query? If the unconscious has contents, such as complexes and archetypes, can probing into contents alleviate the otherness of the unconscious? Are we left with a transcendental style argument or can phenomenology open out the unconscious? A transcendental argument goes from the observed to the unobserved, whereas phenomenology seeks multiple modes of the co-given (in unhiddeness). What are the limits to phenomenology? What is the scope of a transcendental argument?

Is there a dimension of *Sophia* or wisdom to nature? Can both the *how* and *where* of wisdom be understood by finite or cosmic consciousness? Can finite consciousness know when it participates in transhuman wisdom? What is the relationship between cosmic consciousness and wisdom? Can the wisdom evolve? How does the wisdom quicken selving? Is the wisdom necessary to involution? Finally, what is the primal question beneath all the others? Perhaps it is the question, "What, in the end, is the nature of nature, or does nature have any nature at all?"

Acknowledgments

As coeditors, Jea Sophia Oh and Joseph E. Harroff, would like to together express their most profound and convivial gratitude to all of the authors included in this volume for their patience and endurance throughout the various stages of this process. We would also like to acknowledge our deep appreciation and sincerest thanks to Robert S. Corrington for his wise words contributing to this volume which once again proves him to be a very creative and challenging interlocutor, and for being most generous with his time for carefully responding to each one of the chapters. Without his philosophy, mentorship, and guidance this book would not have been possible. We also would like to thank Marilynn Lawrence for graciously contributing one of her amazingly beautiful paintings of natural landscapes for the cover design of this book. And we offer our profound appreciation to the publisher, Lexington Books, and especially to Jana Hodges-Kluck, for all of her steadfast efforts, insightful suggestions, and persevering energy and to Sydney Wedbush and Arun Rajakumar for carefully poring over the entire manuscript and coordinating an optimizing editorial process seeing this book to the presses. And finally, we the coeditors would like to here recognize each other as beloved and intimate "other selves" (*jakiyah* 자기야; 自己) seeing all this through to the end—sine qua non.

Introduction

Joseph E. Harroff and Jea Sophia Oh

> The community of interpreters is fragile, open ended, hermeneutically sophisticated, eschatologically open, non-tribal, non-violent, rich in novel interpretants, not self-absorbed but other directed, pluralistic, and capable of compassion.
>
> —Robert S. Corrington[1]

There is nothing outside of nature while nothing is purely natural. This volume is a collective outcome of our shared process of becoming together as a community of interpreters who have perhaps been rendered even more vulnerable by attempting to be unveiling the shadowy sides of nature and even the gaping "holes," while attempting to philosophize in the face of tragic suffering and radical evil in order to become collective agents of healing cultures. This is no naively optimistic task, in that the perspectives articulated herein tend to abjure any ontological absolute or ethical perfectionism. Instead, we take seriously the tragic historical and cultural limitations placed on even the boldest imaginative visions of future-oriented "beloved community." Though perhaps a bit unrealistically, we have decided to participate in a symbiotic selving and god-ing process together in order to imaginatively transcend the "fate" that what we are now in given the apocalyptic ecological catastrophes and political turmoil causing a rapid increase of human suffering around the world. Despite all the entanglements of bad ideologies, cruelty and hate, we continue to hope for healing nature-culture's manifestations of "sacred folds." Robert Corrington states that "nature's folds come and go outside of any principle of sufficient reason or any structure of intelligibility . . . they are beyond good and evil" and "the emergence of good and evil traits within the human order."[2]

Alfred North Whitehead once diagnosed "misplaced concreteness" as a pervasive category mistake wherein ontological discreteness is assumed to stand in for complex entanglements of historically inflected yet indeterminate vital-material processes.[3] With a similar set of concerns John Dewey referred to the all-too-frequent tendency to take the identified consummatory results of a historical process and reify them into a single causal principle overarching everything at the beginning of a process as some sort of static and totalizing *telos* as "*the* philosophical fallacy" (emphasis ours).[4] Both Dewey and Whitehead presented diagnosis of a reductionistic and distorting tendency in thinking, and we can approach the philosophies of pragmatism and process as marking a sea change in philosophical activity wherein complex systems and social organization replace foundational individualism and atomized subjectivity formation. Say what you will about the tradition of substance ontology and atomized discreteness in the history of Western philosophy (for the Rationalists and Empiricists alike), one thing is certain—namely, the "problem of individuation," that is, how a thing, process, person, or event is differentiated from the unsummed totality of all other "things," in the broadest sense of the term, is actually quite easy to deal with if we posit foundational subjects in the form of "soul stuff" or a *res cogitans* standing behind the phenomenal changes. But much like the old adage that "all is right with the transcendent God in heaven," the relatively simple problem of individuation marked at core by a kind of ontological transcendence (of complex relationality and historicist materiality) becomes quite messy and muddled when we begin to address axiological considerations and how it is that wholly transcendent beings might be held responsible for the presence of radical evil. So it is with some sense of relief perhaps that mainstream theodicies of old are seemingly no longer relevant to a metaphysics of process and relational ontologies of experience. But it is not clear that the so-called problem of Evil doesn't still present a deep challenge for any post-metaphysical or post-theological attempts to think through the role of ideals (aesthetic, ethical, political, and religious values) in shaping our attempts to sustain fragile intergenerational and intercultural projects of communicating communities of interpretation.

Robert Corrington's hermeneutic horizon of ecstatic naturalism, as a new chapter in America's pluralizing philosophical culture, moves beyond both of these aforementioned reductive and metaphysical fallacies by positing deep values—divinity is a relational property of nature naturing ecstatic processes and consummatory events. The old Aristotelian divisions between aesthetics, ethics, and religious experience are not applicable to a stance of radical "ontological parity" between all natural complexes in an infinite order of nature as an ontologically abysmal ground yet phenomenologically rich tapestry of potentially consummatory becoming. We might say that for Corrington the greatest error in thinking stems from egomaniacal and fascistic "characterological armorings" that attempt to locate a unified theory of everything under the banner of an absolute trait that all experience homogenously possesses. In rejecting any reductionistic naturalisms and totalizing materialisms Corrington avers to the radical pluralism at the heart of his pragmatic orientation that moves away from Absolutism toward a more conversational and convivial experimentalism:

> The philosophical idea of "matter" is hopelessly confused and has little role to play as the foundational category. In fact, naturalism, in its ecstatic mode, affirms that there can be no *one* trait found in each order of nature. Every order has just the indefinite number of traits that it has and these are ramified in and as nature. Philosophers seem addicted to the quest for just such *a* universal trait. A few examples of such an imperial move are actual occasions, monads, substance, sense data, ideas in the mind of god, proper names, spirit (*Geist*), the dialectic, forms, organism, and internal relations (emphasis in original).[5]

Any pragmatic prophecy then, speaking from the ontological difference, can only invoke so many "lures for feeling" as always contingent and fragile cues for communicating communities to come together more expansively and ecstatically, as *imagined* (but for that reason no less real) communities practicing creative democracy. This is a *democracy-to-come* wherein ecstatic interpretants are all striving to realize relationally constituted *potentials* as contributing to an always provisional and fragile achievement of consummatory goods and future-oriented ideals in biotic and semiotic cultures of value.

The very ideas of good and evil, historically conceived in Western philosophy primarily as axiological absolutes, would seem to be just a particularly dramatic instance of misplaced concreteness in a world of contingent becoming and fragile cultures of values and valuing. Indeed, the onto-theological world picture of a wholly transcendent divine monarch in need of a theodicy from his mundane and messy creatures, due in large part to the glaring logical inconsistency and ethical contradiction between omnipotence and omnibenevolence in a world rife with undeserved suffering, moral tragedy, and heinous acts of senseless evil, would seem to be an increasingly obsolete relic of a particularly bad brand of metaphysical theologizing that divorces Reality/Appearance, Being/Becoming, Transcendence/Immanence, God/Nature, and so on. As a horizon of interpretation that neither posits a transcendent order of metaphysical value nor rejects the divine possibilities of nature naturing to achieve creative ethical and aesthetic advance, ecstatic naturalism is a most fecund discursive framework that can be appealed to in engaging with actual "problems of evil" like police brutality, racist-sexist-ableist-classist violence and all forms of cruelty, nuclear proliferation, and anthropogenic climate chaos in a world of pure becoming daily facing structural injustices and tragic violence all the while teetering on the actual brink of apocalypse and still haunted by an imperialist Cold War geopolitics that always threatens total conflagration.

The chapters in this volume singly and together then should be approached as so many unique experiments in thinking through the implications of an ecstatic naturalist horizon of interpretation in addressing the place of values that sustain cultures of healing and the radical hope necessary for cultivating moral and political imaginations of democracy-to-come as an always horizonal limit concept or regulative ideal that actually intervenes in the flesh and blood of relationally constituted human persons as they emerge with and for each other in families and societies working against any authoritarian inertias that permit the continuation of evil structures of domination. By "healing cultures" we intend to foreground the significance of the active, gerundive, and processual nature of an immanent transcendence or "ecstatic naturalism" as a creative horizon for realizing values of intersubjective flourishing, while also highlighting the significance of culture as an always unfinished project of making interpretive and ethical space available for the voice of the voiceless and the not-yet-heard demands made upon the heart-minds of those feeling the push and pull of the total fabric of the cosmos and the ecstatically emergent communicating communities of interpretation as the "starry heavens above and the moral law within;" but this time not just rationally determined, but taking into full account the "wild" ethical and political imagination capable of being embodied by those willing to risk letting down their authoritarian tendencies and fascist prejudices as so much "characterological armoring" while putting on the faith of archetypes of healing that require truth and reconciliation between oppressor and oppressed, victimizer and victim of violence and trauma, and let loose the radical hope of healing through suffering in a ceaseless communicating community of creatively democratic interpretation.

CHAPTER THEMES

In "Providence and Providingness: On Platonic and Ecstatic Naturalist Good, Evil, and Infinity" Marilynn Lawrence offers an account of how Stoic, Platonic, and early Christian philosophers, though diverging on important details, all accepted a notion of a divine providence (*pronoia*) governing and ordering the world. Creatively drawing upon Corrington's ecstatic naturalism that rejects traditional supernatural or transcendent providence in favor of Buchlerian "providingness" or natural grace, Lawrence establishes an ontological difference "between natural complexes and providingness." Tracing how Plotinus and later Platonists such as Proclus inherited from the Stoics and Middle Platonists the problem concerning the causal efficacy of divine providence given the existence of evil, she articulates how these Neoplatonists describe the workings of providence through their theories of nature, while putting distance between the demiurgic god and the source of evil without resorting to Manichaean or Gnostic dualism. In Plotinus, the middle Platonic division of providence into primary, secondary, and tertiary becomes two types of providence: upper and lower. Lawrence's chapter critically examines how secondary providence and the Neoplatonic lower providence are horizontal or circular rather than vertically causal. Together with tertiary providence (daimonic providence), these forms more closely support ecstatic naturalism's ideas of nature's sustaining providingness as a form of the infinite—that is, as a god-ing activity wherein natural grace and grace of the spirits are on par with each other. By abandoning the vertical and transcendent primary providence, ecstatic naturalism is left with the important task of providing an alternative, naturalistic approach to ethics without relying upon absolutistic conceptions of good and evil. Toward this task, Lawrence highlights often overlooked aspects of ancient neoplatonic thinking, such as the interworking of lower providence, fate, and daimons. In doing so she offers many historical and philosophical resources for reimagining the relationship of the psyche to nature and the divine.

In "The Experience of Values and the Possibility of Ordinal Phenomenology in Corrington's *Deep Pantheism*" J. Edward Hackett provides a comprehensive overview of the phenomenological tenets of what Corrington calls *ordinal phenomenology* and attempts to tease out how this phenomenological backing supports Corrington's distinction between nature naturing (*natura naturans*) and nature natured (*natura naturata*). Offering eleven characteristics of ordinal phenomenology and treating them like axiomatic propositions, Hackett focuses his argument on the claim that Corrington need not invoke phenomenology for what speculative metaphysics can do for him already. In arguing that since nature naturing as an unconscious is something that cannot be given and thus cannot be supported using phenomenology, Hackett wants to show that nature-as-given conceals within itself innumerable orders, and the fact that they are given out of the hidden potential of nature's unconscious cannot do away with certain problems of emergentism. Hackett's ultimate aim in this chapter is to cast doubt on a seemingly central tenet of ecstatic naturalism, namely that there can be no absolute ontological othering of divinity; instead he wants to suggest that classical theism or panentheism are logically plausible accounts despite Corrington's best efforts to employ "ordinal phenomenology" to bracket bad metaphysics. In suggesting that ordinal phenomenology does not prioritize any specific interpretation of nature, Hackett intimates that the radical phenomenological method might have to presuppose the very things it wants to say are provided in experience as insight. This is a complex claim to unpack and it might overlook many of the psychoanalytic contributions to Corrington's ordinal phenomenological method for approaching experience in its multitudinous sublimity. Perhaps a return to the things themselves via William James's "radical empiricism" wherein nothing more or less than the blooming buzzing confusion of experience is taken in toto could further help problematize Hackett's contention. The chapter concludes by considering some interesting conceptual distinctions drawing from a personalist lineage of philosophical literature and wondering whether or not deep pantheism represents a kind of impersonalism insofar as it seems to deprivilege the personalist and phenomenological primacy of the first-person perspective as the only way to encounter values.

In "Dwelling with the Deep Ones: Lovecraftian Horror and the Selving Process" Thomas Millary takes up H. P. Lovecraft's science-fiction horror stories as constituting an important entry in the fabric of American literary experience, and presents a creative reading of this fiction as involved in a philosophical enterprise of engaging the sublime terror of nature's infinity. Millary argues that a nuanced psychoanalytic reading of Lovecraft can help discern the philosophical significance of his work, and that Robert Corrington's method of ordinal psychoanalysis can provide us a lens to understand the relationship between Lovecraft's unhappy life, his racist social perspectives, and the metaphysical insights that may be gleaned from his stories. Corrington's philosophy of ecstatic naturalism and Lovecraftian fiction both attempt to decenter the human subject in intriguing ways and both describe the shattering experience of recognizing nature's boundless indifference. Since, as Millary suggests, we can find a more complex and varied perspective on the powers and potencies of nature in ecstatic naturalism, we can find fecund resources for engaging with Lovecraft's fiction beyond the seeming nihilism of experiencing the forces of nature as exclusively horror-inducing. By considering Lovecraft's personal and philosophical commitments to rigid regularity and stern consistency through Corrington's approach to how birth trauma inflects the selving process, Millary does much to clarify the origins, values, and shortcomings of this corpus of imaginative fiction.

In "On Being Sunk?" Desmond Coleman engages with W. E. B. Dubois's famous ventriloquizing of a variant of a question perpetually thrown at him by progressive whites of his time: "How does it feel like to be a problem?" Instead of approaching this question through the arguably more familiar metaphor of the Veil and the "color line" in American experience, Coleman explores this vital question through the philosophical and filmic metaphors of "sunkenness" and "the Sunken Place." The former, "sunkenness," is a concept-metaphor used by G. W. F. Hegel throughout his oeuvre to address the determination of the "mental and spiritual characteristics" of the "Negro race," while the metaphor of "the Sunken Place" is deployed by writer-director Jordan Peele in his critically acclaimed worldwide blockbuster *Get*

Out. While these two creative geniuses are thinking centuries apart, and while it is unlikely that Peele borrowed this concept-metaphor from a reading of Hegel, the resonances and dissonances of their use of the notion of being sunk vis-à-vis the experienced meaning of blackness raises important existential and ethico-political questions not only about the logical and historical conditions of possibility regarding this conceptual connection—between "blackness" and "sunkenness"—but also about the fraught relationship between this lived connection and the experience of those people called "black." Coleman takes this connection into the historically uncanny realm of the alchemical in tracing a genealogical connection between elemental thinking in Western and non-Western (Arabic and African philosophizing) thought traditions, while providing a rich reading of Peele's filmic imagination and its relationship to the individual and collective unconscious as it relates to the ecstatically natural tropes of blackness, sunkenness, and salvific experience.

In "Racism, Religious Education, and Transformation" Moon Son and Ji Young Park cast important theoretical and practical light upon the emergent field of transformative ecology and religious education to dismantle racial discrimination and cultural prejudice as highlighted by the inequities and tragic injustices brought into stark relief by the Covid-19/SARS-2 pandemic. The threats of ethnic and racial discrimination like white supremacy and cultural imperialism have been used to foster cultural division and fear of alterity which at root is a fundamental disregarding of the rich tapestry of global cultural heterogeneity and its pluralistic potentiality for cosmopolitical conviviality. Even though the traditional sign systems of yin-yang 陰陽 dipolarity in Eastern thought have the contrasting characteristics and productive relational tension, they can provide meaningful possibilities to transcend any reified state of static being and rather invoke in experience a call to be ecstatically transforming with and for each other in ethically edifying interpretive community. Son and Park's creative chapter in this newly emergent field of ecologically aware comparative religious education is rooted in recognizing an ecstatic transformation occurring between *nature naturing* and *nature natured*; a transformation to be discovered in the communicative consciousness of an always relational and culturally contextual selving process that can transcend any static reification of self and otherness. This fascinating sample of a broad-ranging intercultural and global conversation to come arrives at the position that our social bodies and co-selving spirits, as severely isolated and quarintined during the pandemic, may prove to have a hermeneutical vaccine value to intensify and accelerate the dismantling of racial discrimination and cultural imperialism on a global scale within an ecstatic transformation of comparative religious education. Such a bold educational worldview can encompass both the carving and polishing of the jade ritual implement of embodied betweenness and can better construct an ideal of transcendental sunyata as radical democratic openness rather than nihilistic emptiness.

In "A Phenomenological Study of Feminist Political Consciousness" Susan Erck quite provocatively foregrounds the idea that women's adherence to patriarchal doctrine(s) and habituated praxis is not simply a matter of being ideologically "duped," but rather stems from a deep-seeded pathology of being experientially and psychologically derivative in a much more phenomenologically complex manner. Following in the tradition of Mary Daly, Erck explores the psychosocial contexts of violence carried out against women through a religious studies framing. Erck's theoretical stance posits women's actions of "victim blaming" to be part of a complex psychoanalytic and phenomenological experience of "characterological armoring" in response to certain triggering and traumatizing experiences part of a broader psychospiritual and somaesthetic mechanism of anxiety abatement. With this critical ethical aim at the heart of her project, Erck mobilizes Robert Corrington's ordinal phenomenology of intra-tribal religious violence, drawing upon the Reichian psychoanalytic concepts of "emotional plague" and "characterological armoring." An important strength of approaching women's complicity with and in patriarchal violence at large by using Corrington's Reichian-inspired terms of "armoring" to cover up and deflect against the terrifying aspects of various "holes in nature" is that such a hermeneutical horizon allows for a more expansive and nuanced examination of precisely when such "armoring" practices might be removed or revoked. It is precisely this dynamic of the selving project and how it relates to fascist and patriarchal violence that Erck exposes in her lucid reading of the formation of feminist consciousness in Christine de Pizan's *The Book of the City of Ladies*. This literary corpus serves as a fecund illustration and trope for reading against the grain the psychological processes and phenomenological framework involved in cajoling, permitting, and enabling women's complicity in patriarchal violence.

In "Recapturing World-Loyalty: A Relational Response to Ecological Violence" Katelynn E. Carver considers the basis of the ethical problem of violence. The perpetuation of violence in any context ethically demands adequately formulating the *why* of that particular act of violence so as to better seek out the *how* of ameliorating its traumatic after effects and ideally working to transformatively reverse it. Carver's chapter offers a robust contextualization of the problem of violence that considers the driving framework of interpretation to be always oriented toward applicability. In expanding the concern with violence beyond any reductive anthropic exceptionalism, Carver ruminates on how the Earth itself is often quintessentially Othered, and she calls our ethical attention to the task of discerning how we might engender familiarity and ultimately affection for the Othered Earth.

To do this she argues we must first do adequate justice to the primordial violence of Othering that has led to the current state of affairs. Moving beyond philosophical and empirical attestations of exclusive humanism in myriad contexts has indicated the enduring power of such violence and its brutalizing consequences for interconnected relationships. Carver suggests that we enter the preexisting majesty of the natural order and participate in the various "ecstasies" on offer as reverent and wonder-filled children of an ecstatically open-ended creative process of *nature naturing*. With such a paradigm shift in interpreting violence we can foster change and redefine the hermeneutic resources we teach the generations to come: and only then, can we hope to have better stories to tell the children.

In "Fecundity and Healing of the Great Mother: Reading Corrington's *Nature and Nothingness* via Yin-Yang Thinking" Jea Sophia Oh reveals how both women and nature are placed at the bottom of an anthropocentric hierarchy and have been collectively subalternized by violent anthropogenic climate changes across global, regional, and local scales. Nevertheless, such victimization of the subaltern not only weakens the power of subalternity but also limits its very conceptual boundary. Oh's comparative reading of Corrington's *Nature and Nothingness* carried out via a dynamic process principle of yin-yang correlative cosmology seeks to provide a healing philosophical framework and a fecund analogy to address this apocalyptic time of rampant species endangerment and mass extinction, the global climate emergency, and the increasing rise of racist and xenophobic right-wing authoritarianism. By no means setting herself an easy task with all of this, Oh introduces the Korean term *salim* (enlivening) and offers it as a resourceful healing spirit amid these pandemic/pandemonic times in which we live. Along with Corrington's notion of the Great Mother, Oh considers how *salim* and other archetypal notions need not be essentialized or statically genderized to serve as productive metaphors for healing cultures to subvert and transform patriarchal violence and anthropic exceptionalism. Despite the genderized degradation of both women and *salim* tasks, it is important to remember that the activities of *salim* make things alive and enliven nature at home in a family and in the world. Hence, *salim* cannot be essentialized as a singular trait of masculinity or femininity but rather is better understood as a dynamic "fecundity." This unleashing power of life exists in nature, in every living and nonliving aspect of nature as a healing energy. Drawing upon resources in Confucian and Daoist cosmologies of relational flourishing, Oh affirms that life itself comes from a hidden energy (yin) which is ceaselessly being made manifest (yang) in nature through the ongoing circulation of the non-dual yet dipolar flows of yin-yang (陰陽).

In "Evil as Human Resistance to the Indifferent Force of the Primal Nature: On Storytelling Animals" Iljoon Park sets out to prove two things:

(1) that evil itself (as a singular or foundational trait of nature) does not exist and (2) that what we call "evil" is anything experienced as a result of the basic human resistance and antipathy to the indifferent forces of primal nature which produces what Corrington labels "holes in nature" and "totalizing nothingness."[6] With these two propositions, Park highlights the narrative importance of evil as it is produced by human interpretive resistance as part of the ubiquitousness of storytelling in human life. In natural communities, marked by historical inertias of identitarian violence, evil tends to get identified as demonized other(s) and manifests as religious, and other forms of extremist violence, othering entire communities of persons under the influence of "emotional plague."[7] Such ruthless violence in natural communities may be a response to one's experience of "holes in nature," which are narratively signified as evil, although this total demonization is not a healthy solution. Park alerts us to the dangers that face natural communities when they cannot come up with satisfactorily acceptable interpretations of what problems and predicaments they are facing (i.e., felt resistance to natural experience). Park thinks that wherein the meaning of nothingness is generally denigrated throughout communities, it may be experienced as the "totalizing nothingness," especially when human beings "have a low tolerance for chaos and ambiguity."[8] In such scenarios the demonized other(s) become the sacrificial elements for the despairing natural community in its felt rage and resistance to "evil." Park concludes this chapter with a plea for telling better stories in communicating communities of interpretation so that we can hope to heal and transform together. Otherwise, all we can expect is a continuation of the mimetic cycles of natural violence wherein communities get stuck in the rut of endlessly demonizing others for their own moral inertias, resistances, and shortcomings.

In "The Posthuman and an *Advaya* Dialectic of Sacrifice" Ick-Sang Shin aims to reveal the meaning of *relationships* in what some have dubbed our posthuman condition. Starting with the assumption that there are but two main ways of defining posthuman subjectivity—one is an anthropocentric view of classical humanism and the other is a post-anthropocentric view of anti-humanism—Shin reveals that despite the important differences between the two definitional stances, both positions agree in understanding the posthuman subject as a relational and interdependent nexus of experience. And according to Shin it is precisely here that a real problem for thinking emerges: even if relationality appears with multiple names, such as interdependence, interconnection, belonging to a multitude, otherness, non-unitary subjectivity, nomadicity, becoming, and so on, are we not at root leaving the basic existential question regarding the specific mechanisms or modalities of relating unexplained? In other words, must relationships be bound up with equal reciprocity, or not? And even if their might be ethically acceptable forms of non-equal reciprocity, how to realize such reciprocity in actual relationships? Posthuman relationality inherits the character of biotic relationships, as long as the posthuman through technological mediation are not completely separated from human elements or still actively imitating human traits. However, since Shin argues that the relationships pertaining to the maintenance of life are often not mutually sustaining or interactive, but are often one-sided acts of energy consumption, we arrive at a logic of sacrifice. Even in posthuman situations then, the sublimity of sacrificial logic will continue. Through examining the homeostasis mechanisms of life Shin shows how the problem of subjectivity in a posthuman condition still inevitably touches upon the problem of forced sacrifice. Moreover, since this mechanism is dialectical, and has an immanent transcendence featured in its temporal structure, Shin proposes that we make creative appeal to a Buddhist conception of non-equal reciprocity within homeostasis mechanisms of life, namely as an embracing of "non-duality" *advaya* (不二的, *bul-ijeog*) as a dialectic of sacrifice. Shin takes the chapter into interesting comparative theological ground by developing the dialectic into a *messianic* sacrifice via Giorgio Agamben's dialectic of double negation, and concludes with a reinvigorated concept of *ecstatic* self-transcendence as a live option for subjectivity formation in the (post) human present.

In "Education for the Symbiosis of Humans and Machines in a Posthuman Age" Eunkyoung Lee considers how as a "posthuman" (or perhaps more aptly phrased "transhuman") age continues to arrive, contemporary developments in science and technology, which have sought to enhance human abilities or even go entirely beyond human limitations, are now capable of dreaming of overcoming the very biological category of "human being" via technological means. Lee shows how the prefix "post-" while not necessarily referencing a biological extinction of human beings, instead referring to the fact that human being, as a fixed ontological category with definitive traits, is becoming increasingly nebulous as it passes through technological mechanisms and innovations. In the process of transforming into a new digitally networked being via artificial intelligence, digital networks, virtual reality, and various other forms of cybernetically augmented reality-constituting technologies, Lee detects a looming crisis of civilization. She warns that as the category of the "human" becomes increasingly unmoored from its biological and historical centers of subjectivity formation, that without an adequate response in cultural and educational resources and strategies, the end result can only be nihilism. Hence, Lee makes a case for the importance of rethinking education, particularly within the humanities, as a vital strategy for ameliorating, or at least postponing, any looming global crisis of cybernetic nihilism. Thus, a most urgent task for the coming age will be comprehensive educational reformation.

In "Selving in a Dangerous World: William James, Buddhism, and Ecstatic Naturalism" Jonathan Weidenbaum asks if it is possible to reconcile the active and individual self with a more profound sense of our inherence within a greater whole? The mission of this chapter is to argue that such a synthesis is offered in ecstatic naturalism's concept of the *Selving process*. The very label denotes a project instead of a soul or substance, and is thus going against the grain of much Western thought on the question of personal identity that tends to focus upon ontic categories of self-same identitarianism. Weidenbaum proceeds by hermeneutically unpacking the meaning of a more process-oriented notion of "selving" with the assistance of one of the venerable ancestors of ecstatic naturalism—William James. Both James's landmark chapter on the self from *The Principles of Psychology* and his later concept of "pure experience" are mined for insights in order to highlight a few of the main features of the Selving process. Weidenbaum then proceeds to carry out an intercultural philosophical inquiry via the early and later James in conversation with Buddhist thinking, with particularly emphasis upon a few of Corrington's most keen statements concerning the Selving process as found in his dialogue with the great third-century Mahayana thinker, Nāgārjuna. In outlining strong similarities between a Buddhist account of personal identity (namely, the notion of *anatta*, or no permanent self) with the selving models offered by James and Corrington, Weidenbaum is able to highlight some important differences between Buddhism and these two American philosophers—particularly in regard to the nature of the environment through which Selving is presumed to occur. James articulates a *concatenated* universe wherein the disconnections between events are every bit as real as their connections—an ontology in which individual things are not, finally, dissolvable into a mere web of relations, and ecstatic naturalism follows along similar lines, and presents a metaphysics in which the *natural complexes* which comprise the cosmos are only unevenly related to one another. While James writes of a *melioristic* universe in which our efforts may bring about the "world's salvation," Corrington describes how an acceptance of the occasionally chaotic, even dangerous nature of reality can play a role in our self-knowledge and personal development. Moreover, the goal of the Selving process is to introduce a measure of freedom and creativity into his or her culture, thus helping in the struggle against authoritarian societies and the push toward more progressive ones. Weidenbaum deftly shows how for both James and Corrington, both James and Corrington, the self is an activity which operates in and through a larger set of relationships—and it is the acknowledgment of the somewhat messy but improvable character of this broader context that permits a sharper form of purpose for human beings.

In "Redemptive Suffering with *Tianming* 天命: An Ecstatically Naturalist Reading of Sacred Selving in Early Confucian Ethics" Joseph E. Harroff articulates certain family resemblances of ethical-political thinking between

Robert Corrington's ecstatic naturalism as a (post)theological stance and early "Confucian" (Ru 儒) philosophical discourse of role ethics (倫理道德) with a non-monotheistic conception of divinity as found primarily in the perspective of the *Mengzi* in the Warring States Period (475–221 BCE). Harroff suggests that we can fruitfully draw from Corrington's ordinal phenomenology and neo-pragmatist perspective that engages in theorizing holes in nature as ecstatically natural complexes to optimally approach a productive interfaith conversation with Si-Meng Confucian role ethical discourse regarding redemptive suffering. The idea of a ritually communicating community acting creatively and intergenerationally to sustain contingent values and living hope for the arc of history to bend toward a more just and sustainable world order is an important trope that Corrington's non-foundational rejection of totalizing metaphysical world pictures can help us approach in coming to understand the unfamiliar assumptions animating the early Confucian conception of persons called to a higher purpose by a heavenly invoked order of nature-culture (*tianming* 天命). Such an intercultural conversation can help in raising vital questions regarding the morally transformative significance of responsibly contextualizing senseless suffering, moral tragedy, and radical evil within historical-cultural frameworks of interpretation that don't require the positing of any theo-ontological absolutes or any wholly transcendent and omnibenevolent forces in nature.

In "We have Met the Enemy, and He Is Us: The Nature of Evil and the Evil of Nature in the Anthropocene" Robert W. King explores the concept of evil in an age marked by the impossibility of not finding the trail of the human serpent cast over everything. According to Jedediah Purdy, the Anthropocene "finds its most radical expression in our acknowledgment that the familiar divide between people and the natural world is no longer useful or accurate."[9] This development problematizes the notion of a purely natural Evil or Suffering separated from the human environment of modern, industrial society. King's chapter aims to further explore the consequences of the Anthropocene concept for considerations of Evil and Suffering inherent in Nature, with considerable attention paid to Ralph Waldo Emerson and William James, both of whom King positions as "Stoic optimists" who in various ways wrestled with the existence of evil and suffering in their intimate worlds. Both Emerson and James wrote passionately about our need to recognize and engage with the evil and suffering in Nature and human experience, rather than turn away in bad faith attempts to find consolations in supposedly transcendent Realities behind appearance, or in the "characterological armorings" involved in the pathologies always haunting the decline of democratic societies and personalities into the malaise of fascist authoritarianism. Moreover, this is a moral imperative that resonates profoundly with ecstatic naturalism, and, in the Anthropocene, becomes ever more vitally imperative.

Suffering and Evil in Nature: Comparative Responses from Ecstatic Naturalism and Healing Cultures is like a beautiful interwoven tapestry of such colorful threads of transdisciplinary and comparative responses to the problems of suffering and evil in nature. We are with Corrington in hoping for ever more wholesome, edifying, and sustainable communities of interpreters as healing cultural agencies operating beyond static and absolutist conceptions of good and evil:

> As this process continues to unfold, I marvel at creativity of the human spirit and its capaciousness when the conditions allow. Above all, it is in furthering the quest for justice and the renewal of democracy that is the most basic goal of ecstatic naturalism.[10]

NOTES

1. Robert S. Corrington, *Nature's Sublime: An Essay in Aesthetic Naturalism* (Lanham: Lexington Books, 2013), 103. Reprinted with permission. All rights reserved.
2. Corrington, *Nature's Religion* (Lanham: Rowman & Littlefield Publishers, 1997), 135.
3. A. N. Whitehead, *Science and the Modern World* (New York: The Free Press, 1925; 1967), 51.
4. John Dewey, *Human Nature and Conduct: An Introduction to Social Psychology* (New York: Henry Holt and Company, 1922), 175.
5. Corrington, "Foreword" in *Nature's Transcendence and Immanence: A Comparative Interdisciplinary Ecstatic Naturalism*, Marilynn Lawrence and Jea Sophia Oh eds. (New York: Lexington Books, 2018), x.
6. Corrington, *Nature and Nothingness: An Essay in Ordinal Phenomenology* (Lanham, MD: Lexington Books, 2017), x.
7. Ibid., 1–2.
8. Ibid., 14.
9. Jedediah Purdy, *After Nature: A Politics for the Anthropocene* (Cambridge: Harvard University Press, 2015), 2–3.
10. Corrington, "Foreword" in Nature's *Transcendence and Immanence*, xiii.

REFERENCES

Corrington, Robert S. "Foreword" in *Nature's Transcendence and Immanence: A Comparative Interdisciplinary Ecstatic Naturalism.* Marilynn Lawrence and Jea Sophia Oh eds. New York: Lexington Books, 2018.

———. *Nature's Religion.* Lanham: Rowman & Littlefield Publishers, 1997.

———. *Nature's Sublime: An Essay in Aesthetic Naturalism*. Lanham: Lexington Books, 2013.

———. *Nature and Nothingness: An Essay in Ordinal Phenomenology*. Lanham, MD: Lexington Books, 2017.

Dewey, John. *Human Nature and Conduct: An Introduction to Social Psychology*. New York: Henry Holt and Company, 1922.

Jedidiah Purdy. *After Nature: A Politics for the Anthropocene*. Cambridge: Harvard University Press, 2015.

Whitehead, A. N. *Science and the Modern World*. New York: The Free Press, 1925; 1967.

Part I

A DEEP OPENING OF NOTHINGNESS

SOME METAPHYSICAL RESOUNDINGS

Chapter 1

Providence and Providingness

On Platonic and Ecstatic Naturalist Good, Evil, and Infinity

Marilynn Lawrence

INTRODUCTION

Approximately two thousand years ago, Stoic philosopher Seneca asked, "If a Providence governs the world, why do many evil things happen to good men?"[1] Two hundred years later, Plotinus reexamined the existence of this governing and ordering principle when he asked, "If the Reason-Principle [*logos*] of the universe is the creator of evil, how is that not injustice?"[2] Stoic and Platonic philosophers took different approaches to the topic of providence's reach and the existence and meaning of evil, but they all accepted a concept of *providentia* or *pronoia*—that is, the foreknowledge or care of a godhead for what follows from it, or, a divine force that governs and orders the world. The scope and meaning of providence, including these two senses, is expressed in numerous ancient treatises on the topic. The notion of divine providence extends at least as far back as fifth century BCE in, for example, the historian Herodotus[3] and the tragedian Euripides,[4] although it wasn't until Plato's treatment of *pronoia* in the *Timaeus* dialogue[5] that it became a philosophical theme picked up by writers of the Hellenistic period and Late Antiquity.[6] Seneca, Philo of Alexandria, Plotinus, and Proclus wrote dedicated treatises on divine providence, and it was a major theme for middle Platonic authors discussed later as well for Christian philosophers such as Clement, Origen, and Augustine.

Returning from the ancient world to the present, Robert Corrington has several times discussed ecstatic naturalism's acceptance of Justus Buchler's concept of "providingness" in place of providence. The ontological difference, or rather "natural difference," has been described as that fissure "between natural complexes and 'providingness.'"[7] This difference reflects the two aspects of nature as nature naturing (which provides the conditions of possibility for natural complexes) and nature natured (the manifestation of natural complexes).[8] The need within ecstatic naturalism to replace or eschew the ancient concept of providence is quite understandable, particularly when this term can imply the intervention of a transcendentally supernatural creator god who impinges upon events and lives in a created natural world. Yet, the various thinkers who grappled with providence in non-Christian Middle Platonism and Neoplatonism contexts still provide us with food for thought, particularly regarding the nuances of how this concept relates to others such as fate, chance, and contingency. Their concepts of providence also have to be squared with the realities of evil, badness, and suffering. A closer look at providence in the Platonism of Late Antiquity can highlight features that relate to and support ideas within the fold of ecstatic naturalism—particularly *sustaining providingness as the infinite*, *god-ing*, *natural grace*, and *grace of the spirits*. At least temporarily, this will require cutting off the supernatural head of providence, so to speak, while leaving the intelligent body. This body is revealed in the phenomenology of natural complexes and the selving process.

THE TRIPLE PROVIDENCE OF MIDDLE PLATONISM

We'll begin the question of providence in an awkward, but vital, stage of philosophy's development, the first few centuries of the Common Era. This was a time when Aristotle had been rediscovered after lying fallow in a period dominated by Stoic, Epicurean, and skeptical thought; when inchoate Christian communities were creatively dealing with a radical event and conception of God, but had not yet settled upon the doctrine; when Philo of Alexandria's Platonic Judaism was making an impact on the philosophical community; and when a wide variety of philosophical and religious ideas were exchanged freely in Alexandria, resulting in creative combinations sometimes deridingly called "syncretism" or "eclecticism" by historians.[9] In this milieu, a handful of Platonists, such as Plutarch, Apuleius, Alcinous, Numenius, and Calcidius (now classified as Middle Platonists in contrast with the skeptical Academy and the "Neoplatonists" to come), worked through the top issues of their day—providence, fate, human choice, chance, the nature and functions of various gods and daimones, the possibilities of divination, and how all of these factors impact the soul.

Several Middle Platonists describe the doctrine of a threefold providence that includes primary, secondary, and tertiary expressions of providence.[10] Much of what we know about this doctrine is through the work *On Fate* attributed to Plutarch.[11] Primary providence is identified with the will and mind of the principle god, which was thought to be the demiurge of Plato's *Timaeus*.[12] We could stop here and this would provide the more common definition of a vertical and remote providence that governs the world—a providence in opposition to Buchler's providingness. However, Middle Platonists were also responding to the Stoics' immanent concept of providence that was intertwined with a deterministic fate. For the Stoics, there is no clear distinction between God, Nature, Logos (rational principle), and Providence—so when they advise us to live "in accordance with nature" (*kata phusin*), they mean to live according to the rational principle or Logos, which also means according to Providence and Fate. Anchored by Plato's discussion of and acceptance of a concept of fate (e.g., *Republic* Book X), the Middle Platonists acknowledged a place for fate but did not believe that all natural events and lives are predetermined by an immutable chain of causality. To preserve the transcendence and goodness of a primary god, they proposed a secondary providence that governs generation and destruction within the natural and temporal world. Secondary providence is that of the "young gods" (*neois theios*) of the *Timaeus* who serve as administrators to the demiurge.[13] They can be identified with the planets and stars of the zodiac, residing along the circles of the Same and the Other.[14] Pseudo-Plutarch says fate is co-born or begotten along with (*suggennêtheisa*) this secondary providence.[15] He describes it as being infinite (apeiron) in some ways and finite in others, for it operates in the motion of a finite circle, and the time that measures this motion is also circular or cyclical.[16] The place of secondary providence is a seat lying just on the edge of nature; for, the visibility/sensibility of the bodies of these younger gods places them partially within the nature that they oversee.

The Middle Platonic association of secondary providence with fate (*heimarmenê*) provides more grounding to understand its unique role in the natural world. Fate itself was understood as "hypothetical" (*ex hupotheôs*), meaning it is identified with the law of actions and consequences that determine the orders of that which is possible (*to dunatos*).[17] Fate, as law, is contrasted with that which is fated (*kath' heimarmenên*), namely, the particular consequences determined by the necessity of this law. That which is possible within hypothetical fate is then split into necessary and contingent factors. Borrowing these divisions from Aristotle, the Middle Platonists further split the contingent into "the usual" (regular occurrences in nature), "the unusual" (spontaneous aberrations from the norm), and things or occurrences which have an even chance of possibility, including human freedom. Chance (*tuchê*), or meaningful coincidence as defined by Aristotle in *Physics* Book 2, is the result of an intersection between human action and the unusual.[18] Necessity, contingency, spontaneity, and chance are all forces present within fate as law or secondary providence, according to this Middle Platonic account.

We might find this secondary providence analogous to the sustaining providingness of nature naturing as "the 'provision and determination' of traits or complexes," as Buchler defines it.[19] Secondary providence is not the personal care of a single creator god for humans and other natural beings, but is the sustaining condition for all potentialities. Nor is it the providence of God's direct intervention in the world, an invisible hand as it were that guides history and individuals, as one might find in dominant forms of Christian theology. Certainly, the story of Christian providence throughout the centuries is a complicated one, but we might contrast the Platonic providences with the thinking of Augustine, who begins *De ordine* (*On Order*) with the question of "how on the one hand, God takes care of human affairs, and on the other these same affairs are shot through with so much evil."[20] Augustine does not want to solve the problem of human perversity by denying God's providence and ordering intelligence from reaching all within nature. Buchler's providingness was conceived as a replacement for the supernatural imposition of a single ordering, rational principle, a purposive order of orders governing the world that we find in Augustine's account—which is in fact the primary premise of Augustine's work *On Order*. Buchler's identification of nature naturing with a provisioning of traits was intended instead to "amplify the suggestion of sheer putting forth, sheer geniture, for better and for worse."[21] Hypothetical fate, as identified with secondary providence, is ordinality, on the edge of that which is generated within nature, or the *natured*. Contingency and chance are included in this ordinality, as are opportunities for both goodness and perversity.

While primary providence is vertical and transcendent, secondary providence is circular, and closer to nature in this Middle Platonic schema. In fact, it defines the possibilities of nature natured. Like primary providence, the secondary is not personally involved in individual lives. For this, the Middle Platonists identified a tertiary providence that is associated with the daimones, a class of beings that concern themselves with specific parts of nature, as well as with human and animal souls.[22] Pseudo-Plutarch describes tertiary providence as the domain "of all the *daimonês* who are stationed in the region of the earth and are guardians and watchers over human affairs."[23] Daimonic providence, along with chance and human choice (literally "what is according to us" [*eph' hemin*]), is contained within fate as law, just as fate is within primary providence. Not much is told by Ps-Plutarch about how this tertiary providence operates within nature, although he does illustrate it with the famous example of the daimonion of Socrates that guided his actions and associations. Ps-Plutarch quotes Plato's *Theages* on this point, where Socrates recounts how his daimonion warned him that Charmides should not set out to train for the Nemean races.[24] Charmides ignored Socrates and was met with a bad fate.[25] This example implies that one can advance in some way through contact with daimonic powers, including the gift of foreknowledge of the consequences of actions within the network of fate, or that which is called "fated" (*kath' heimarmenên*). We'll find more clues about the meaning of the immanent daimonic providence in Neoplatonic developments of the topic.

THE LOWER PROVIDENCE OF NEOPLATONISM

The trifold providence of Pseudo-Plutarch became divided in Neoplatonism into an upper and lower providence, where the lower form functioned like secondary and tertiary providence combined. Both third-century Plotinus and fifth-century Proclus discussed providence in the context of the traditional problems of evil. In Neoplatonism, providence can only be good, for it is sourced in the Good, which is identified with the One. The various arguments on providence made by Plotinus and Proclus are too numerous and complex to outline here. It may help then to simply note that their positions on providence and evil are alike in some core ways, such as in their exoneration of the higher principles of One and Nous from any direct responsibility for genuine evil. Their philosophical idiosyncrasies on providence reflect their overall differences in metaphysical schemata and distinctive interpretative stances regarding Plato's texts.

For present purposes, our focus is on a few instances of lower or immanent providence, and how it may resonate with concepts elaborated within ecstatic naturalism. However, one can immediately see that providence's association with a transcendent Good is a stumbling block in these comparisons because the locus of Buchler's and Corrington's providingness is beyond good and evil. Corrington identifies Buchler's providingness with the concept of the "sustaining infinite," which he describes as follows:

> It obtains prior to the distinction between good and evil, and prior to any axiological distinctions such as those aesthetic distinctions so prized in process form of naturalism. The sustaining infinite sustains the just and the unjust, the beautiful and the demonic, the fragmented and the harmonious, the honorific and the detestable, the living and the dead (via effects), and the realms of the possible and the actual.[26]

Corrington also describes his related concept of natural grace as "beyond good and evil as its moral upshot may be genuinely ambiguous from the standpoint of human goods and values."[27]

Nevertheless, in this comparison between providingness and lower providence, we can retain from Neoplatonic providence the notions of goodness or the beneficial, as well as the conditions for possibilities (as suggested in the earlier discussion of secondary providence), while momentarily leaving behind an external, ultimate source outside of nature. We can also compare how Neoplatonism and ecstatic naturalism both offer ways to make sense of phenomena in the natural world that are experienced as divine; and how the human psyche experiences things as either good or bad.

Providence, for Plotinus, is unified in its essence, but divided in its function. He writes, "There is one providence; but it is 'fate' beginning from the lower level; the upper is providence alone."[28] The function of upper providence is to maintain the unity of what he calls the "true cosmos" (*alêthinos kosmos*),[29] which is essentially the noetic "intelligible nature" from which "second nature" (*phusis deutera*)—our natural world—is derived. One activity of providence is to establish the order and laws of nature in general. Our physical cosmos is not a true unity, but a mixture of competing principles and forces pulled together in a song with various notes and harmonies.[30] For this metaphor Plotinus draws upon the myth of Book X of Plato's *Republic*, where each Siren located on her own sphere emits a unique tone that contributes to the music of the spheres. The idea is that components within nature (complexes) each have their own note or function that are unified into a whole by an immanent rational principle that is a trace of the Good/One. As noted earlier, fate, co-born with secondary providence, was not a predetermined destiny but a law of necessity. It is hypothetical such that if certain things happen, certain consequences follow. Fate itself appears to be value neutral (neither good nor bad), while providence does convey a thriving goodness even if the beings it touches are not equipped to appropriate all that it may provide on account of their own capacity.[31] Plotinus often points out that humans are not the best beings in the universe, but have a mixed nature, sitting between gods and beasts; yet, they have a share of wisdom, intellect, skill, and justice.[32] And, as he writes, "something may shine in [them] as the stars shine in the heaven of the gods."[33]

A kind of immanent providence is also metaphorically described by Plotinus as follows:

> The workmanship which produces wonders in rich variety in ordinary living beings and the beauty of appearance which extends to the fruits and even the leaves of plants, and their beauty of flower which comes so effortless, and their delicate variety.[34]

Genetic variety and natural beauty are praised as gifts of this lower providence, which also provides natural healing such as cellular repair: "Health is given by an organism's providence [*kata pronoian tou zôou*], when a cut or injury of any kind occurs, the directing rational principle again afterwards joins it and closes the wounds and sets right the suffering part."[35] This function of providence can be thought of as the provisions guiding nature's will to life. The thread of goodness within nature, for Plotinus, extends down to the lowest things of Earth—each has will or striving for the next level of intelligence. Sensible matter seeks to be formed into living matter, which then strives to be elevated to the higher part of the world soul—the Soul moving to Nous (Mind or Intellect) and ultimately to the One.[36]

If providence touches beings in the natural world with goodness, identified with divinity in Platonism, what of evil or badness? At the most general level, evil has no independent being of its own for Plotinus, but is an absence of good that emanates from the One to Nous to Soul to Nature. What philosophers call "natural evil" in contrast with "moral evil," namely natural disasters, viruses, death of organisms, and so on, is not actually bad in the larger scheme of things, for the world can't have generation without decay and destruction.[37] Animals preying upon each other is simply an aspect of necessity, and of the cosmos as a whole organism. Neither providence nor its source is to blame for natural evil. Plotinus tells us that if we were to demand that natural evil be abolished, we would be demanding providence to be abolished—there's no point to providence's existence without multiplicity.[38]

Natural evil (as apparent evil) and the providence that works through it are part of the natural order of things. The relationship of providence and evil is made even more complicated when Plotinus discusses moral troubles that pertain to human souls. In one sense, there is an acceptance of the human condition. Just as we don't blame plants for not having higher cognitive capacities, we shouldn't blame humans for not being gods.[39] On the other hand, training the soul toward wisdom and other virtues can result in its return to more divine circumstances as the ascent of the soul.[40] While providence is always operating in the background as a linking and unifying thread of intelligence, we can either act in concord with it (*sumphôna*) or not—our actions are not *caused* by it.[41] Human wrongdoing is not caused by a metaphysical principle of evil,[42] but is a consequence of our actions and our ignorance about what is within our responsibility. It occurs when the human soul forgets its true nature as it is caught in a web of diverse phenomena, concerns, and destructive passions and beliefs.

Human evil or corruption of goodness can be contrasted with Plotinus's conception of metaphysical evil, which is *nonbeing*, a total lack of measure and limit as the infinite or indeterminate (*apeiria*).[43] Evil is associated with the infinite on account of the absence of any unifying or limiting power to act upon multiplicity. This absolute unlimited is also identified by Plotinus as matter-in-itself (i.e., not sensible matter that has taken shape or form).[44]

Matter is en-formed and becomes sensible through the trace of the goodness of the One into the depths of nature. On the point of evil as infinite nonbeing, an intriguing comparison could be made between this aspect of Neoplatonism and Corrington's *infinite* as a property of "naturing nothingness," and more specifically its mode of "totalizing nothingness" as described in *Nature and Nothingness*.[45] In ecstatic naturalist terms, the human being has access to this form of nothingness through the feeling of horror in the face of nature's indifference and the threat of dissolution of unity and integrity guiding the human process.[46] While Neoplatonic evil is not described in phenomenological terms, one can imagine the anxiety in an encounter with formless "matter" devoid of all qualities and meaning.

Remarkably, though, the infinite appears at both ends of the Neoplatonic metaphysical spectrum—the absolute Good (the One) and absolute lack of good and being (nothingness as abyss).[47] Plotinus occasionally uses infinite (*apeiria*) to describe the power of the supreme principle, the One. For instance, in a thought exercise, he asks his listeners to imagine the visible universe and all creatures in it, then a sphere that contains all things, then take away all mass and place, and all gods which are powers of this whole, then arrive at this "entire power, extending to infinity and powerful to infinity; and this is that which has become so great that its parts have become infinite."[48] Plotinus recognizes, however, that the infinite cannot be conceived:

> When you come up close to it and do not throw any limit over it like a net you still have it slipping away from you and will not find it in any one thing. . . . But if you approach it as one, it will appear many; and if you say that it is many, you will be wrong again: for if each [part] of it is not one, all of them cannot be many.[49]

Proclus, a fifth-century Neoplatonist, also speaks of the infinite in his work *Ten Problems Concerning Providence*. For Proclus providence from the One is both unitary (*henoeidous*) and of infinite power, which he describes as "more infinite" (*apeiroteron*).[50] Perhaps more infinite is meant to contrast the One's infinity with that of metaphysical matter/evil. This description may aid us in understanding the "sustaining infinite" that Corrington associates with providingness.

Just as the Neoplatonic One is beyond Being, naturing naturing is prior to Being, and the natural difference, the fissure between naturing naturing and nature natured, is prior to Heidegger's ontological difference.[51] When we return to this gap, this abyss between the two modes of nature, we encounter the "sustaining infinite" that quietly makes orders and traits within nature possible. As Corrington writes:

> Another term for "sustaining" is "providingness" (Buchler) that conveys not providence or an alleged divine hand, but the sheer thereness of what prevails in the way it prevails. . . . Perhaps it is best to say that the sustaining infinite is that which transforms nothingness in its various manifestations.[52]

As providingness has been compared here to secondary providence, the hypothetical fate of Late Antiquity's Platonism, the concept of sustaining infinity gives a dimension to this activity, as the potency to shape or en-form the second infinite of nonbeing, of multiplicity or foreboding matter in the raw.

ADDITIONAL CONSIDERATIONS FOR ECSTATIC NATURALISM

There are many potential considerations in exploring ancient ideas to build a robust and perennial ecstatic naturalism, such as Neoplatonic ideas of providence, good, and evil. Some notions will naturally be rejected, while others may hint at a similar framework, language, or interpretations of nature and the human process within it. Suggested earlier is an association between the circular secondary and lower providence with providingness, for it is neither a vertical principle of divine intervention, nor predestined plan, but the providing and sustaining conditions for the natural world—an activity of nature naturing. The infinite in Neoplatonism also supports the ideas of naturing nothingness in its modes of totalizing, sustaining, and encompassing. By excluding supernatural primary providence and its sources in the One and Nous, new questions arise concerning the source and nature of the Good, or that which is beneficial for all of nature and human beings within the purview of ecstatic naturalism. One approach may be to reframe the goodness conveyed by Neoplatonic providence as Wisdom, which Corrington describes as a power of the Great Mother. He identifies this archetype as "a healer, a goad to artistic greatness through an ongoing transit between chaos and creativity, a counselor, a visionary, and a witness to one's inner heart."[53] Another way is to understand the One not as supernatural but as an ancient expression of the same ineffability that is at the limit of human comprehension—which can be identified with one Nature in its two modes or aspects.[54]

We might also consider what the activities of an immanent and natural providence would look like. Tertiary or daimonic providence appears to drop out of Plotinus's account of providence, but this does not mean that Plotinus didn't have a philosophical account of the daimon. In fact, he dedicated an entire treatise to this topic (namely, *Ennead* III.4). Furthermore, the role of daimones in the context of providence is more directly addressed by Proclus, who extended the more immanent and personal sort of daimonic providence to angels and heroes. He claimed that sometimes providence reaches directly below these classes of being and sometimes it operates through them, for they have knowledge and foresight beyond temporality.[55] Daimones and gods were thought to communicate through divination and dreams.[56] This was a commonplace in the ancient Greek–speaking world, but dream divination and dream therapy were of particular philosophical importance for Neoplatonists such as Porphyry, Iamblichus, and Proclus.[57] In response to questions posed by Porphyry on dream divination and "god-sent" dreams (*theopemptoi oneiroi*), Iamblichus makes note of the different phenomena experienced by the soul as it separates from the body in intellectual contemplation and in sleep.[58] Divination through dreams obtains different levels of clarity based on the preparedness and connection of the dreamer's soul to higher noetic powers. At these clearer states, the soul has knowledge of the ordering of the cosmos, and even participates in the order, care, and improvement of affairs including, Iamblichus tells us, the healing of sick bodies through dream therapy.[59] Proclus also refers to dream therapy (*oneirôn therapeiais*), divination, and visitations of gods and daimones as gifts of providence.[60]

Phenomenologically speaking, ancient philosophers who attempted to integrate the ontology of daimones and angels into their metaphysics were working with phenomena that we still experience today, and still may choose to interpret as divine inspiration, an inner voice, a higher self, the Muses, and so on. Concerning gods and daimones communicating through dreams, from an ecstatic naturalist perspective, such archetypal "big dreams" or "deep dreams" that are perceived as coming from a divine source contain ejects of nature naturing that aid the selving process through access to nature's unconscious.[61]

This cohort of Platonists from Late Antiquity paid special attention to the "guardian *daimôn*" described by Plato as a being that is selected by the soul descending to earth (*Rep.* 617E), and which plays out in his dialogues through the references Socrates makes to his daimonion (e.g., *Euthyphro* 3B, *Apology* 31C, *Rep.* 496C). We can think of the guardian daimon as a "witness to one's inner heart" or earthbound emissaries of the Great Mother. Taking cue from Carl Gustav Jung, for Corrington, this archetype offers both the potential for healing and chaotic destruction.[62] However, the Great Mother is perhaps too encompassing of an archetype and requires assistants who bring more personal manifestations of the archetypal ejections from nature naturing—daimones, and the guardian daimon in particular, who provide caring wisdom throughout significant moments in the life of an individual. The daimonic, however, also carries with it the potential for the *demonic* in the negative sense. As Corrington notes in *Nature and Spirit*, "Any given epiphany of power may be both demonic and divine in its manifestation."[63] In ancient Greek terms, the numinous is *deinos*: both awesome and dreadful.

The numinous is perceived as demonic when posing a threat to the self. However, as long as it carries hope with it, "the numinous object empowers the self to go beyond previous assimilative strategies and to find new means by and through which to approach the divine."[64]

In our look at circular or horizontal providence, we can also consider Corrington's distinction between natural grace and grace of the spirits. Perhaps natural grace, described in the postscript to *Deep Pantheism* as "a gifting of the potencies of nature to the human process," is also akin to the secondary providence of Middle Platonism.[65] Grace of the spirits, that is, "plural in its manifestations and appears through the god-ing energies that burst upon the Selving process," may be what the ancient Platonists considered as manifestations of tertiary or daimonic providence.[66] Another potential reading would be to relate the sustaining and "worldhood making" secondary providence to Emersonian emanations and tertiary providence to the Schellingian ejected energy.[67] At the very least, there is a similar recognition in both Late Antiquity and ecstatic naturalism of different but related experiences that suggest these pairings that describe processes of a psyche's relationship with nature.

A final consideration is that of reincarnation of souls. In *Nature and Nothingness*, Robert Corrington begins a tentative discussion of karma and reincarnation from an ecstatic naturalist perspective.[68] He notes that ecstatic naturalism does not entail belief in reincarnation, although it is a concept that can be reframed within naturalistic metaphysics.[69] The Western version of reincarnation as education of souls is part and parcel of the Platonic tradition inherited through the Pythagoreans and Orphic mysteries. This belief involves a combination of necessity, freedom, chance, and a guardian daimon. Neoplatonists sought to reconcile the cycle of rebirths with the human experiences of injustice and suffering. Proclus, for example, addresses the problem of how suffering for actions in past lives that one does not remember can advance the soul.[70] He argues that wrongdoers who are aware of their crimes and punished for them typically do not cease from perpetuating the same crimes. While providence works as a healer of souls it does not always work quickly. Swift retribution or punishment does not always serve as the appropriate medicine. The Platonic view of karma does not satisfy those seeking vengeance for wrongs committed, but operates case by case in its own propitious moments (*kata kairon*) according to the means that would best improve the soul.[71] In a similar vein, Corrington speculates that the soul forgets past incarnations because:

> The current psyche simply cannot handle the overload that would come from allowing past lives to become conscious. . . . The "invisible hand" of nature protects the self from its own unconscious overload that could even produce madness were it unleashed.[72]

For both Proclus and Corrington, an immanent providence or natural grace protects the health of the soul from overwhelming recollections.

The theosophical tradition that Corrington draws upon for his thoughts on reincarnation, represented by Helena Blavatsky and Annie Besant, was heavily influenced by a late nineteenth-century revival of Neoplatonism prompted by the English translations of Thomas Taylor and Alexander Wilder's passionate syncretism of Neoplatonism and Hinduism.[73] Further research into Neoplatonic writings may help flesh out the questions of how a naturalistic conception of reincarnation can be philosophically conceived within ecstatic naturalism, or how polytheistic and henotheistic perspectives are compatible with deep pantheism. Corrington makes gestures in this direction in his most recent book *Nature and Nothingness*, where he poses questions about the status of gods and goddesses of various religious traditions within an ecstatic naturalist perspective—are they archetypes or real entities? He notes, "Thus far we can say that the gods and goddesses who are prevalent in many world religions prevail somewhere between the domain of unconscious projections and that of literal beings."[74] Perhaps, from a pragmatic and phenomenological stance, this is asking the wrong question. Archetypes can be encountered as real beings and can be valuable to selving in such an encounter. Corrington postulates that archetypes are the genus of which deities are the species, playing a role in human evolution, evolving or dying out.[75] This fine line of polytheism and deep pantheism can be informed by the approach taking by the Neoplatonists, especially that of Proclus for whom the traditional Greek gods and goddesses were preserved in philosophical theory and practice by being both real entities and metaphysical principles at the same time. The gods Kronos and Ares (their significances by the time of Proclus were filtered through the astrological doctrine), for instance, are gods that function in the world as principles of cohesion (*sunochês*) and separation (*diairesiôs*).[76] Perhaps drawing from such traditions can help us to identify the significance of numinous ejections from nature's unconscious.

NOTES

1. *De providentia*, I.1. Seneca begins his essay with this question that was posed by his friend Lucilius: *Quaesisti a me, Lucili, quid ita, si providentia mundus regeretur, multa bonis viris mala acciderent.*
2. *Ennead* III.2.10. The A. H. Armstrong edition and translation is used unless noted otherwise. *Plotinus: Enneads*, Loeb Classical Library, (Cambridge, MA: Harvard University Press, 1966–1988).
3. Herodotus. *Historiae* 1.62, 3.108.
4. Euripides. *Phoenissae* 637.
5. Plato. *Timaeus* 30C, 44C.

6. *Pronoia* was also associated with the goddess Athena as attested to by Demosthenes (*In Aristogitonem* 1.34) and second-century CE Pausanius (*Graeciae descriptio* 10.8.6). However, rather than Forethought Athena, her epithet may have been originally *Athena Pronaia* (literally, Athena in front of the temple). See James Frazer's commentary to his translation of Pausanias for details. *Pausanias's Description of Greece*, vol. 5, ed. and trans. J. G. Frazer (Cambridge University Press, 1898), 251–252.

7. Robert S. Corrington. "Ordinality and the Divine Natures," in *Nature's Perspectives: Prospects for Ordinal Metaphysics*, eds. Armen Marsoobian, Kathleen Wallace, and Robert S. Corrington, Albany (NY: State University of New York Press, 1991), 359.

8. Ibid.

9. See *The Question of "Eclecticism": Studies in Later Greek Philosophy*, eds. J. Dillon and A. A. Long (Berkeley: University of California Press, 1988).

10. This doctrine is found in Pseudo-Plutarch, Apuleius, and is described by fourth-century bishop Nemesius. Calcidius offers similar views on fate and providence in his commentary on the *Timaeus*, but discusses a single providence and attributes the functions of secondary providence solely to fate, as the "decree of providence." His discussion of daimones as caretakers of human beings also functions like tertiary providence.

11. Pseudo-Plutarch, *De fato*, in *Plutarch: Moralia*, vol. 7, eds. and trans. Phillip de Lacy and Benedict Einarson (Cambridge: MA, Harvard University Press, 1959), 303–359. The current author is of the opinion that this work, which does not have the polish of authentic works by Plutarch, is an outline or notes taken from a lost work that may or may not have been by Plutarch.

12. *De fato*, 572F. Ps-Plutarch is drawing upon the demiurge of the *Timaeus* as the primary god in his discussion of fate and providence. However, the status of the demiurge among Plato's interpreters, including Neoplatonists, is widely debated. The issue is whether the demiurge represents the hypostasis Nous or something between Nous and Soul. It does not represent what would be the closest concept of a supreme God in Neoplatonism—the One. For a clear account of these positions, see the dissertation by Jason G. Rheins, "The Intelligible Creator-God and the Intelligent Soul of the Cosmos in Plato's Theology and Metaphysics" (2010), *Publicly Accessible Penn Dissertations*. 184.

13. Plato. *Timaeus*, 42D.

14. For more on this topic, see M. Lawrence, "The Young Gods: The Stars and Planets in Platonic Treatment of Fate," in *Perspectives sur le neoplatonisme*, eds. M. Achard, W. J. Hankey, J-M. Narbonne (Quebec: Presses de l'Université Laval, 2009), 95–110.

15. Ps-Plutarch. *De fato*, 574B.

16. Ibid., 569C.

17. Ibid., 570A-B.

18. For a preliminary discussion of the role of Aristotle's divisions of chance and spontaneity in later Platonic metaphysics, see M. Lawrence, "The Place of Chance or Fortune in Platonic Fate," in *Conversations Platonic and Neoplatonic: Intellect,*

Soul, and Nature, eds. John F. Finamore and Robert M. Berchman, Sankt Augustin, Germany: Academia Verlag, 2010, 87–101.

19. Justus Buchler, *Metaphysics of Natural Complexes, Second, Expanded Edition* (Albany, NY: State University of New York Press, 1990), 200. Cf. 3, 100, 276.

20. Augustine. *De ordine*, 1.1. *quam quomodo fiat ut et Deus human curat et tanta in humanis rebus perversitas usquequaque diffusa sit*, trans. Silvano Burrosi. *St. Augustine: On Order* (South Bend, IN: St. Augustine's Press, 2007), 3.

21. Buchler. *Metaphysics of Natural Complexes*, 276.

22. This word is left untranslated throughout because, while this class of beings could be mischievous, they were typically not demons or instigators of moral evil, as later portrayed in the Christian tradition. In the Platonic tradition, daimones are ontologically superior to human beings. For example, in *Ennead* VI.7.6, Plotinus describes them in the chain of being as between gods and human beings, where a daimon is dependent on a god, as a human being is dependent on a daimon, particularly the guardian daimon that is assigned to a soul which is born human.

23. *De fato*, 574a. Translation by R. Sharples and P. J. van der Eijk in *Nemesius: On the Nature of Man* (Liverpool: Liverpool University Press, 2008), 209, n. 1009.

24. Ps-Plutarch. *De fato*. 574B-C.

25. The authenticity of *Theages* has been debated by modern scholars, although it was accepted as authored by Plato in antiquity.

26. Corrington. *A Semiotic Theory of Theology and Philosophy* (Cambridge: Cambridge University Press, 2000), 111.

27. Ibid., 19.

28. *Enn*. III.3.5.15–17.

29. Ibid. 2.2.

30. Ibid. 2.2.28–32.

31. Ibid. 3.3.

32. Ibid. 3.9.

33. Ibid. 3.14.

34. Ibid. 2.13.

35. *Enn*. III.3.5, A. H. Armstrong translation, slightly modified by the author.

36. For example, *Enn*. 6.7.25.24–26. "For form is the good for matter, for if it had sense-perception, it would welcome it, and soul [is the good] for body—for without it, it could not exist and be preserved."

37. *Enn*. III.2.4.

38. Ibid. 3.7.

39. Ibid. 3.3.

40. Following Plato's metaphysics of the descent of individual souls to earth, for Plotinus and later Neoplatonists, the individual soul descends from its home in the noetic realm (or rather, a part of the soul descends while the true self or soul always remains in the intelligible world), and the soul has the capacity to return or make the ascent through philosophical training (and theurgical training for Neoplatonists such as Iamblichus and Proclus). See, for instance, *Enn*. IV.8.

41. *Enn*. III.3.5.

42. Dominic O'Meara. *Plotinus: An Introduction to the Enneads* (Oxford: Clarendon Press, 1993), 84.

43. *Enn.* 1.8.6; 6.6.1.

44. For example, Ibid. II.4.15; I.8.3; I.8.13–14.

45. Corrington. *Nature and Nothingness: An Essay in Ordinal Phenomenology* (Lanham, MD: Lexington Books, 2017).

46. For example, Corrington writes, "To experience horror is to encounter the total eclipse of the human process and to sink into permeating sense of nature's total indifference and primal uncaring aspect." *Nature and Nothingness*, 27.

47. Consider that Corrington speaks of two types of infinite, the infinite reach of the two aspects of nature (naturing and natured) and the encompassing nothingness (*Nature and Nothingness*, xi). These may correlate to these two Neoplatonic infinities, should we recognize that the One's infinite power meets resistance of nothingness (whether also produced by the one or independent) in order to manifest anything intelligible or material.

48. *Enn.* V.8.9.24–28. Author's translation.

49. Ibid. VI.6.3.33–38. Trans. A. H. Armstrong. Plotinus. *Enneads* VI.6-9 (Cambridge, MA: Harvard University Press, 1988), 17–19.

50. Proclus. *De decem dubitationibus circa providentiam*, 11.1–7. *Proclus: Ten Problems Concerning Providence*, trans. Jan Opsomer and Carlos Steel (London: Bloomsbury, 2012).

51. Corrington, *Nature and Nothingness*, 116.

52. Ibid., 57.

53. Ibid., 120. Cf. 23.

54. This is not as far-fetched as it seems, given that Corrington has acknowledged the influence of the Neoplatonic One on his thinking, although filtered through Emerson. See his *Deep Pantheism: Toward a New Transcendentalism* (Lanham, MD: Lexington Books, 2016), 12–13.

55. Proclus. *De decem dubitationibus circa providentiam*, 10.65.

56. Ibid., 7.46. For more on daimones and divination, see Sarah Iles Johnston's *Ancient Greek Divination* (Chichester: Wiley-Blackwell, 2008), 9–12, 46.

57. A good resource on treatment of dreams in this time period is Patricia Cox Miller's *Dreams in Late Antiquity: Studies in the Imagination of a Culture* (Princeton: Princeton University Press, 1994).

58. Iamblichus. *De mysteriis* 3.2–4. *On the Mysteries*, trans. Emma C. Clarke, John M. Dillon, and Jackson P. Herschbell (Atlanta, GA: Society of Biblical Literature, 2003).

59. Ibid., 3.3.

60. Proclus. *De decem dubitationibus*, 46.12–15.

61. Corrington. *Deep Pantheism*, 61–62. Also, *Nature and Nothingness*, 64–65.

62. *Nature and Nothingness*, 119–120.

63. Corrington. *Nature and Spirit* (New York, NY: Fordham University Press, 1992), 168.

64. Ibid.

65. Corrington. *Deep Pantheism*, 102.

66. Ibid.

67. See *Naturing and Nothingness*, 116–117.

68. Corrington. *Nature and Nothingness*, 69–76.

69. See, for instance, *Nature and Nothingness*, 69–76.

70. In *De decem dubitationibus*, problem 8 is dedicated to addressing the criticism that providence (through fate) doles out retributions long after the wrongdoing, and this includes retribution for actions in past lives that are not remembered by the incarnated soul.

71. *De decem dubitationibus*, 51.

72. Corrington, *Nature and Nothingness*, 115.

73. For more on the history of Neoplatonism's reception in theosophy, see Jay Bregman, "Neoplatonizing Gnosticism and Gnosticizing Neoplatonism in the 'American Baroque,'" in *Gnosticism, Platonism and the Late Ancient World*, eds. Kevin Corrigan and Tuomas Rasimus (Leiden and Boston: Brill, 2013), 609–629.

74. *Nature and Nothingness*, 121.

75. Ibid., 123.

76. Proclus. *In Platonism Timaeum commentaria*, III.67.22–26.

REFERENCES

Armstrong, A. H., ed. and trans. *Plotinus: Enneads*, Loeb Classical Library, Cambridge, MA: Harvard University Press, 1966–1988.

Bregman, Jay. "Neoplatonizing Gnosticism and Gnosticizing Neoplatonism in the 'American Baroque.'" In *Gnosticism, Platonism and the Late Ancient World*. Edited by Kevin Corrigan and Tuomas Rasimus, Leiden & Boston: Brill, 2013.

Buchler, Justus. *Metaphysics of Natural* Complexes, Second, Expanded Edition. Albany, NY: State University of New York Press, 1990.

Burrosi, Silvano, trans. *St. Augustine: On Order*. South Bend, IN: St. Augustine's Press, 2007.

Corrington, Robert S. *Deep Pantheism: Toward a New Transcendentalism*. Lanham, MD: Lexington Books, 2016.

———. *Nature and Nothingness*: *An Essay in Ordinal Phenomenology*. Lanham, MD: Lexington Books, 2017.

———. "Ordinality and the Divine Natures." In *Nature's Perspectives: Prospects for Ordinal Metaphysics*. Edited by Armen Marsoobian, Kathleen Wallace, and Robert S. Corrington. Albany, NY: State University of New York Press, 1991.

De Lacy, Phillip, and Einarson, Benedict, eds. and trans. *Plutarch: Moralia*, vol. 7. Cambridge: MA, Harvard University Press, 1959.

Dillon, John M. and Long, A. A., eds. *The Question of 'Eclecticism': Studies in Later Greek Philosophy*. Berkeley: University of California Press, 1988.

Johnston, Sarah Iles. *Ancient Greek Divination*. Chichester: Wiley-Blackwell, 2008.

Lawrence, Marilynn. "The Place of Chance or Fortune in Platonic Fate." In *Conversations Platonic and Neoplatonic: Intellect, Soul, and Nature*. Edited by John F. Finamore and Robert M. Berchman, 87–101. Sankt Augustin, Germany: Academia Verlag, 2010.

———. "The Young Gods: The Stars and Planets in Platonic Treatment of Fate." In *Perspectives sur le neoplatonisme*. Edited by M. Achard, W. J. Hankey, J. M. Narbonne, 95–110. Quebec: Presses de l'Université Laval, 2009.

Magee, John, ed. and trans. *Calcidius: On Plato's* Timaeus. Dumbarton Oaks Medieval Library, Cambridge, MA: Harvard University Press, 2016.

O'Meara, Dominic. *Plotinus: An Introduction to the Enneads*. Oxford: Clarendon Press, 1993.

Opsomer, Jan and Steel, Carlos, eds. and trans. *Proclus: Ten Problems Concerning Providence*. London: Bloomsbury, 2012.

Plato. *Complete Works*. Edited by Cooper, John M. and Hutchinson, D. S. Indianapolis and Cambridge: Hackett Publishing Company, 1997.

Plotinus. *Enneades*. Edited by Henry, P. and Schwyzer, H, R., *Plotini opera, 3 vols*. Leiden: Brill, 1:1951; 2:1959; 3:1973.

Proclus. *De decem dubitationibus circa providentiam*. Edited by Boese, H. *Procli Diadochi tria Opuscula*. Berlin: De Gruyter, 1960.

Rheins, Jason G. "The Intelligible Creator-God and the Intelligent Soul of the Cosmos in Plato's Theology and Metaphysics." *Publicly Accessible Penn Dissertations*, 2010.

Seneca. *De providentia*. Edited and translated by Basore, John W. *Seneca: Moral Essays, Vol. I*. Cambridge, MA: Harvard University Press, 1928.

Sharples R. and van der Eijk, P. J., eds. and trans. *Nemesius: On the Nature of Man*. Liverpool: Liverpool University Press, 2008.

Chapter 2

The Experience of Values and the Possibility of Ordinal Phenomenology in Corrington's *Deep Pantheism*

J. Edward Hackett

INTRODUCTION

In this chapter, I propose to review the phenomenological tenets of what Corrington calls *ordinal phenomenology* and tease out how this phenomenology supports his conception of a "nature naturing" (*natura naturans*) and "nature natured" (*natura naturata*) distinction in his book *Deep Pantheism* (2016).[1] To do this I engage in a close reading of the eleven characteristics of ordinal phenomenology and treat them like analytic propositions. In some cases, I will be problematizing the individual propositions, but my main contention starts in the fourth proposition in which I argue that Corrington need not invoke a claim to be doing ordinal phenomenology since metaphysical speculation can supplant the role he thinks phenomenology is playing for him already.

My main argument is that to theorize that nature naturing has an unconscious is a theoretical stance that cannot be given to experience and thus cannot be supported using methods of phenomenology. On the contrary, nature-as-given conceals itself within the innumerable orders, yet the fact that they are given out of the hidden potential of nature's unconscious cannot do away with the problem of Alexanderian emergentism. Therefore, classical theism or panentheism are still logically plausible accounts despite Corrington's best efforts to employ "ordinal phenomenology" to get beyond such limiting paradigms. In this way, ordinal phenomenology does not secure any specific interpretation of nature, but falls prey to the danger of presupposing the very propositions it wants to say are given in experience as insight.

In the second part of this chapter, I assess where we might end up once we give up on the role phenomenology purportedly must play in Robert

Corrington's deep pantheism. Deep pantheism takes on the problematic features of denying the personalist and phenomenological primacy of the first person in which one would encounter values. This makes deep pantheism a form of impersonalism. Impersonalism is the view that elements of the first person are features more generally of some mind-independent (personal-independent) feature. In saying that deep pantheism is a form of impersonalism, Corrington goes through great efforts to depersonalize the metaphors he borrows from persons that he projects onto nature.

THE STATED PROBLEM

An ordinal phenomenology (and phenomenological more generally) cannot do away with the various hierarchies and layers that a predetermined commitment to ordinality and ordinal psychoanalysis allegedly filters since hierarchies reflect the various ways intentional feeling operates a tergo. The really-real emerges into every greater complexity to the point that nature is not simply all there is, but a phenomenology of nature would pick out the hierarchies as they are given to us phenomenologically. Insofar as Corrington desires what I might call a flat ontology of nature, he should abandon phenomenology altogether and on a deeper level, he should recognize that the innumerable order generated out of nature naturing and nature natured give rise to inevitable hierarchies, including the fact that metaphors employed to understand nature are entirely personalistic. Thus, the stronger conclusion of this chapter is that ordinal phenomenology assessed through Husserl as a starting point may be all but impossible, and ordinal phenomenology lends itself to a personalistic conception of nature, not the ordinal conception of nature Corrington thinks he is advancing in his project of ecstatic naturalism. For these reasons, ordinal phenomenology cannot support the metaphysical project of ecstatic naturalism in the way Corrington thinks it can.[2]

THE PROBLEM WITH ORDINAL PHENOMENOLOGY UNDERWAY

First, Corrington says ordinal phenomenology "questions the notion of the evident by arguing that it is too loaded toward the pure and certain."[3] Corrington is aiming these comments at Edmund Husserl, and while Husserl aims at bracketing our presuppositions to get clear on what our relation of intentional acts and objects (the co-given) is revealed in phenomenological description, the nature of the evidence thus disclosed at least after *Ideas Pertaining to a Pure Phenomenology vol. 1* is reducing evidence to the truth of what's revealed. Corrington feels that "bracketing still privileges the internal features of subjectivity."[4] In this way, Corrington is surely right.

Early Husserl did emphasize the constituting features of intentional acts in relation to objects, but clearly Husserl developed other analysis just the same. The fact that we describe intentional acts does not mean we lose sight of the co-given. Husserlian approaches to phenomenology as a method is a loaded field, yet Corrington does not spell out exactly how the evidence of ordinal phenomenology is to be weighed. Instead, Corrington tends to assert that it supports his account of "deep pantheism" as an experiential horizon. In the same way, Husserl and Corrington encounter the same impasse in two different fundamental ways. They both presuppose ontological commitments that later are asserted as evident, though this is a pervasive problem with phenomenology more generally.

Phenomenologists have focused so much on present experience as to what is excluded: each phenomenologist endorses a different metaphysics than the other.[5] In other words, phenomenology presupposes an uncritical metaphysics. In seeking to describe phenomenon as they are given, whatever the chosen target of description, the phenomenologist in question will reify and take confidence in their description to the point they think the insights reached in description primitively basic and applicable to all other experiences. This fault of phenomenology is on the one end of a spectrum. On the other end of the same spectrum lurks seeing what is primitively basic to your phenomenological description by presupposing it: reaching your insight or your bias into what you are supposed to describe. In this way, phenomenology may often transgress the bracketing Husserl fervently defended and made part of his method. When this presupposing the given occurs, phenomenologists legitimate their own biases.

Second, ordinal phenomenology deepens the notion of "co-given beyond that of portrait painting and the quest for a 'fulfilled' intuition to affirm that all orders of relevance are indefinitely complex and that there can never be a resting place for phenomenological description."[6] While as Anthony Steinbock claims that phenomenological description is for opening "eidetic seeing" or Scheler's "spiritual seeing" as I tend to think of phenomenology, phenomenological descriptions are always in tension with the fact they may be revised.[7] For this reason, phenomenological descriptions seek to disclose insight, and for that function, they are shared with others in order to invite criticism whether or not such eidetic or spiritual seeing occurs. As such, there's no "resting place" within phenomenological method, and this resting place should not be confused with the fact that some phenomenologists, like Husserl and Heidegger, are overly confident in their descriptions. The fact of the matter is that no phenomenological description is free from rest as if someone should just stop and accept phenomenological descriptions without the aim of possibly revising them or wholly replacing incomplete descriptions for more thorough ones. This is true whether or not phenomenology is operating on the static level of discerning the structural features of a perceptual act in relation to an object, or temporalizing the co-constituting relationship as it emerges in the lived experience of history in a genetic phenomenology.

Third, Corrington challenges "the notion of 'phenomenon' as often as privileging spatio-temporal complexes and primacy of sight."[8] In this characterization, some types of Husserl's early phenomenology dealt primarily with perception and therefore sight. Initially, like in his *Ideas of Phenomenology* (1908), phenomenology was static and centered around getting clear on the very epistemic relations we have with the world. However, the fact is that many phenomena are put to the test in Husserl as well as others to warrant the fact that a perceptual-experiential *fringe* (to speak in Jamesian terms) always stands as an irreducible more, or an exceeding otherness, such that phenomena like (ethical, aesthetic, and religious) values are equally described in phenomenology in the same vein. For instance, values deriving from the emotions are secondary in Husserl to a purportedly more foundational epistemic orientation in experiencing them. For this reason, Corrington might be overstating his case against Husserl (and maybe phenomenology more generally) when he suggests that ontological parity should replace Husserl's alleged privileging "of the Ego, universal in its structure and ways of acting, over everything else" and that "Husserl's blatant metaphysical hierarchy clearly violates the spirit of parity."[9] Allow me to explain.

Corrington has embraced a metaphysical reading of Husserl while not paying attention to the ontological neutrality of his phenomenological descriptions and at the very least if that ideal should be governing phenomenological descriptions in general. In order to do phenomenology, one must be provisionally committed to describing elements of experience as they are undergone by the first person. One engages in the epoche, and then in suspending any presupposition to what will be revealed in the elements of experience, which include both the noetic acts and noematic object. Thus, to lay claim to the purely descriptive enterprise then one is already natural with respect to what is being described. To claim, then, that one should put the intentional relation on par with the world in general then reinscribes the neutrality of phenomenological description with a commitment it never possessed procedurally. More to the point, ordinal phenomenology then presupposes the ontological parity it wishes to prove in ordinal phenomenology. I speculate this is due in part by a belief that no such phenomenological description could escape a hermeneutic background of the inquiry, so Corrington may not be bothered by presupposing what is already there to be found. Or perhaps, while we may be tempted to regard Husserl suspiciously through the lens of Paul Ricoeur's famous idealist interpretations of his efforts through which many encounter Husserl, Corrington offers no primary passages to motivate this reading despite its interpretive importance in situating ordinal phenomenology in relationship to his project. Instead, Corrington seems to assert that both ordinal and classical phenomenology have "different metaphysical commitments," yet I am pressing Corrington's approach here in seeking more than an interpretive assertion. Such an interpretive move requires explication that we never get in *Deep Pantheism*.

On the one hand, Husserl's phenomenology may develop into a form of idealism if and only if Husserl never abandons the attempt at describing as neutrally as we can what is self-given. Corrington should make good at the very least that neutrality was never possible, and at least in *Deep Pantheism*, the attempt to show that neutrality was never possible is not done. The idealist interpretation, however, cannot be imposed upon Husserl as scholarly convention insists without first distorting the aims of phenomenological description itself in which the world and its objects are just as needed as conscious acts that intend them.[10]

Fourth, Corrington insists that "all self-showing must be filtered through a process of ordinal psychoanalysis." Self-showing phenomena enter the field of phenomenological attention, and then they are described within whatever limits of attention tied to the particular phenomenologist. However, nothing comes from the unconscious in phenomenology since it requires a "ray of regard." Phenomenology, then, actively pays attention to conscious acts and the intentionality of those conscious acts. If we become aware of the unconscious, then this must act, as Corrington posits, via "a filter." Given nature's self-transforming capacity to reach beyond itself ecstatically, ordinal phenomenology catches these unconscious flinging dynamics from nature like Merleau-Ponty's sparks off a fire, and describes them when these unconscious dynamics become manifest. At least, that's the claim. However, at this point, I am wondering what it means to still cling to the term "phenomenology" and what, if anything, Corrington is doing by insisting upon ordinal phenomenology's involvement with the unconscious, a realm that for purposes of all phenomenological description cannot be touched by phenomenological description.[11]

For Corrington, "ordinal phenomenology is the method of deep pantheism because it uses the ordinal concepts" where ordinality designates for him that everything described comes from nature and that nature is all there is. These ordinal concepts, then, presuppose "the frame within which deep pantheism thrives and functions."[12] In other words, phenomenology becomes a way for Corrington to describe what he presupposes as ordinality. At this point, there's been so much presupposing that phenomenological description does not distance itself from the self-showing of phenomena to be of any use to him. Phenomenological description is supposed to reveal the unconcealed and naive assumptions that must be brought to light by rigorous descriptive efforts. In this way, there must be as much distance of the element of that description from what will eventually become its metaphysical interpretation. At this point, Corrington might just offer independent reasons for why one should adopt his ecstatic naturalism on its own merits rather than building up the view with phenomenology. For phenomenology, then, to have any use, it must (at the very least) gain some distance in describing what went unseen. If not, then in effect, phenomenology very well could be shoring up the theoretical biases for ordinality while insisting upon the self-evidence of those biases as the way things are.

So why phenomenology? Why should "ordinal phenomenology" be a "first philosophy" and "so key to deep pantheism?"[13] My hunch is that this temptation to phenomenology rests on thinking through his long attended project of ecstatic naturalism through these thinkers and in doing so, it would seem, that ecstatic naturalism finds support in both what phenomenology as an open-ended methodology does not do and what it can make claims to do. What's more, phenomenology cannot provide evidence and reasons for these speculations given the lack of distance between qualifying what can be given, and then claiming to see what has been presupposed. In my own research into the value theory of Max Scheler I dive into the complexity of tradition to mine resources for my speculation about the ontological status of values to show (at the very least) some minimal consistency with others, so I'm also prone to such adventurous writings not grounded in careful textual support and historical context. In this way, Corrington offers us an ecstatic naturalism and deep pantheism in light of a dialectical tension between a constellation of thinkers, including revising phenomenology. I am inclined to think he does not need to include phenomenologists as far as any claim to phenomenology is concerned with respect to his project. He could just as well jettison any claim to phenomenology, and speculate pragmatically about what he takes to exist and draw out the conceivable effects of what those commitments entail. In any event, regardless of the phenomenology, Corrington's views are not logically inconsistent in any of the metaphysical claims he arrives at in relationship to Husserl's thinking (or Heidegger's for that matter).

Fifth, ordinal phenomenology "affirms that the Selving process is vastly richer and more ordinally complex than either the transcendental Ego or Dasein."[14] The emphasis here on ordinality seems, at best, a want for a naturalism of some variety to enter back into either a hermeneutic and existential description, which is devoid of even making commitments apart from seeing experience as a structure of being-in-the-worldness, or the more idealistic phenomenology of the later Husserl. In addition, there is no unconscious or psychic drives manifesting in the unconcealed depths of the ego. Corrington is committed to the fact that the selving process had to wait for something like ordinal psychoanalysis to reach its full articulation.[15] At this point, why not abandon the claim to phenomenology all together? Indeed, for the rest of *Deep Pantheism*, phenomenology is mentioned briefly but only as a complement to ordinal psychoanalysis, which is the real theoretical articulation of the remaining chapters beyond the Introduction and first two chapters.

Admittedly, Corrington seems to want a language of process, experiencing to belong both to the dynamics of human experience, which he calls the "selving process," but also the element of our unconscious in nature. If that's the case, then Corrington should recast his efforts as a form of philosophical anthropology that improves upon the defects of Husserl and Heidegger, which he is doing anyway by calling attention to phenomenology's lack of attention to the unconscious realm of human experience. However, again, I have to ask if ordinal phenomenology should be rather thought of as a leaping off point rather than as a complement to Corrington's deeper philosophical motives? What is given in intentionality enters the field of conscious attention for description in phenomenology. The fact that nature's unconscious comes from the assumed depths beyond phenomenology's power to articulate indicates that ordinal psychoanalysis has more theoretical importance for Corrington than ordinal phenomenology.

Sixth, ordinal phenomenology "affirms the liberating power of the renewable practice of ontological parity as superior to the *epoché*."[16] Ontological parity means that everything is on equal ontological footing, "everything whatsoever . . . is an order within the one nature that there is."[17] Under an idealist interpretation, Husserl may be suggesting that consciousness matters more than what is appearing (or even more the idealist that all we can know and exist is fundamentally consciousness and mentality) and that the ontology of the *Sinn* or sense of meaning acquires its significance and elevation as being reduced to the uncritical importance of consciousness. The want on Corrington's part is the avoidance of metaphysical hierarchies that reproduce a concealment of nature to show itself as it is. Even the term "nature" for Corrington is but a placeholder for all that there ever could be, again presupposing the ground of what he would want to be described. Ontological parity is presupposed before any phenomenological description, and so Corrington will constrain the given to what he assumes can be given. This want leads us to the seventh feature of ordinal phenomenology.

Seventh, ordinal phenomenology accepts and makes the distinction between nature naturing and nature natured. Both dovetail each other, and ordinal phenomenology would accept that on the nature-naturing side, nature creates itself out of itself and on the nature-natured side consist "the manifest orders of the world, as a fissioning within the one nature that there is."[18] Noting both sides as they function differently leads to the jointly themed eighth feature.

Ordinal phenomenology adds "that the co-given operates differently depending on whether it is dealing with nature-naturing or nature-natured."[19] The former "shades off into fathomable depths of the unconsciousness of nature" whereas the latter co-given of nature natured, the orders of the world, "must attempt to make unhidden traits that are more available than those pertinent to nature-naturing."[20]

A PRAGMATIC JUMPING-OFF POINT?

With the use of such phrases concerning the given, I am reminded of the short section in William James's *A Pluralistic Universe* in which metaphysics is revealed as employing the explicit use of metaphors for God. While not directly important here, James also reveals the power of metaphysics has a great deal to do with understanding the implicit metaphors and figurative expressions that foreground and anticipate the construction of a given philosophical system. If James is right, metaphysics could just very well be the poetics of metaphor employed for pragmatic speculation. In this way, metaphysics is more about speculating the parts we cannot see or directly experience that make sense of experience by the use of analogical speculation. It comes from an aesthetic impulse to connect what is not connected, and we employ a variety of metaphors. Accordingly, Corrington might be drawing metaphors from both the active generative meanings of sensemaking in the meaning-bestowal of consciousness in phenomenology and the resources of Jungian depth psychology. William James writes,

> The first thing to notice is this, that the only material we have at our disposal for making a picture of the whole world is supplied by the various portions of that world of which we have already had experience. We can invent no new forms of conception, applicable to the whole exclusively, and not suggested originally by the parts. All philosophers, accordingly, have conceived of the whole world after the analogy of some particular feature of it which has particularly captivated their attention. Thus, the theists take their cue from manufacture, the pantheists from growth. For one man, the world is like a thought or a grammatical sentence in which a thought is expressed. For such a philosopher, the whole must logically be prior to the parts; for letters would never have been invented without syllables to spell, or syllables without words to utter.[21]

For James, many metaphysicians interpret the whole prior to the parts by assertion. In claiming to do that, however, we notice that they can only talk of the whole through analogy of some part of experience. Certainly, many might agree in attempting to discern ecstatic naturalism metaphors of growth are everywhere, and those metaphors of growth include the unconsciousness of nature from which we cannot really know anything about, but we can assume its ever accruing presence in our own growth. We experience the likelihood of our own unconscious funneling into our waking life. We experience these parts and project them back onto nature to arrive at nature naturing and nature natured. Isn't Corrington simply projecting the personalistic elements of experience onto nature in the very same constraint that James remarked about those that claim to be doing metaphysic? Might this also suggest the reasons why he finds the metaphors of relationality and our own phenomenological intentionality as metaphors to draw from when explaining nature naturing and nature natured?

The phenomenological person is implicit within Corrington's project. Analogies are made to the unconscious and the waking life of the intentionality. As personalism is a philosophy that assumes the primacy of the entire person in terms of ethics and metaphysics, personalism may be a good substitute for phenomenology to draw upon. Depending on the personalist in question, one may also see it as a form of pragmatic speculation, as in the case of Brightman's employment of radical empiricism from William James and offer a way to conceive of the Divine and nature together compatible with Corrington's appropriation of phenomenology and psychoanalysis.

THE PHENOMENOLOGICAL METAPHORS OF NATURE AND THE IMPLICITNESS OF THE PERSON

In advancing my claims, it sounds like the generative power of the intentional act and object relationship of classical phenomenology is read into deep pantheism (as nature herself). This point might be a motivating feature of why Corrington insists upon ordinal phenomenology as a support for deep pantheism, and why he is drawing attention to how each side of the ontological divide behaves conceptually. Corrington is reading both the active level of consciousness intending objects and generating meaning into nature naturing and nature natured. However, a full conception of the person requires the added unconscious of nature, which he is developing as an ordinal psychoanalysis. On the nature-naturing side, Corrington repeats this mysterious depth with words like "hidden potencies," "unconsciousness," and the above "shades off into unfathomable depths." In one way, the insistence upon the act side as that which is hidden from how the objects emerge as conditioned into ordered complexes reveals again that no defense is ever offered for this repeated claim, but also hints at an explicit reading of the phenomenological person and psychoanalytic subject back into both metaphysical categories of nature. These two sides of nature are all there is.

In truth, Corrington is offering us a speculation, and constantly mentioning it as tied to ordinal phenomenology because, I think, it's fair to say he is reading intentionality of classical phenomenology into a personalized conception of nature that never emerges out of itself into something higher. Like a person, there's the active level of conscious experience open to phenomenological description, and the ever more complicated unconscious depth. The insistence of these elements in nature is, I think, evidence for a proto-personalism in his thought about nature. In effect, if his methods are phenomenological, and if he rejects my criticism to keep phenomenology around, then we must ask, "Would remaining committed to ordinal phenomenology be but a way to see the personal in the very fissioning between nature naturing and nature natured?" At the very least, the personal is embedded in the metaphors he articulates about the impersonalized nature rendering nature to be curiously more personal. While Corrington resisted this criticism, the fact is that with ordinal phenomenology, remnants of the personal contaminate claim to ontological parity.

We can press this Jamesian point about personalism further. What is the difference that makes a difference for Corrington's project by trying to articulate deep pantheism with a method of phenomenology? The difference that makes a difference is that we can get our way of experiencing the world on par with what is really there, if what is really there can be given to us in experience. On this point, I have to admit a strong agreement. Our best metaphysical interpretations of reality must mesh with what we experience. In this way, we must then admit that Corrington's deep pantheism and ecstatic naturalism are but one possible interpretation of the universe. A Jamesian will readily admit our inability to tell from a theoretical viewpoint which worldview is more true in terms of having access to the way reality truly is independently of our experiencing it. Instead, the Jamesian makes a decisive move against all metaphysics: "In every genuine metaphysical debate some practical issue, however, conjectural and remote, is involved."[22] The practical issue here is just how much on par with reality is Corrington's metaphysical speculations with actual experience of nature and more importantly if Corrington can regard his efforts as a speculation. Going deeper, what does phenomenology itself offer Corrington? If Corrington can show that what is picked up on in phenomenological description has the same ontological weight as he thinks of nature naturing and nature natured, then he finds legitimation within experience. The problem here is the distance between what is given and what is thought to be revealed in that givenness.

UNRAVELING ORDINAL PHENOMENOLOGY

If there were something higher reached in phenomenological description beyond nature, then Corrington would be in agreement with classical phenomenology about the existence of essences that are arrived at as primitively basic and constitutive of all other meanings of phenomena. What's more, if one allows for a transcendental phenomenology to arrive at the constitutive senses of meaning achieved in intentionality, then one would be guilty of introducing a hierarchy, and therefore a foundation for nature (and any philosophical conception of it). Corrington is consistent and rejects both. In the ninth feature, ordinal phenomenology "rejects the very notion of essence and the practice of *Wesenschau* or showing of the essences of the order," and in his tenth feature, ordinal phenomenology also "rejects foundations of any kind" since "there are no essences, only the traits of varying degrees of scope and instantiation."[23] For him, we cannot privilege any trait or degree of manifestation a trait may possess. Nature may change, may manifest differently tomorrow, and we must be careful to not blind ourselves to that possibility. The universe can grow and change.

The move to phenomenology though is starting to unravel. When we look at our descriptions of experience, we are trying to find some truth by adopting a phenomenological perspective. Phenomenology, pragmatically speaking, is almost always adopted as a metaphilosophical commitment because the natural attitude or some other theoretical feature has explained away the role of the subject in her experiencing the world. Phenomenology is thus a way to return back to the subject and prioritize first-personal experience. In so doing, phenomenology doesn't avoid the worry of hierarchies because as a methodology it seeks to recover personal experience as the very starting point for doing philosophy in the first place. This unnecessarily so then burdens any claim about hierarchy avoidance since all phenomenologies privilege some relational aspect that goes unnoticed elsewhere whether it is the ready-to-hand practical involvements of Heidegger's *being-in-the-world* or Merleau-Ponty's bodily intentionality.

Ordinal phenomenology can only describe traits as they emerge in this fissioning between nature's acts and nature's objects. However again, I must maintain that these generative powers are at the heart of what it means to be a *person*. For the purposes of my argument, it doesn't matter if person is understood as transcendental ego, Scheler's usage of person, or a Jamesian relationality. The fact is that person is undergirding the metaphor that Corrington does not acknowledge lurking in his explanation of the two sides of nature. For this reason, we can say that the person is the part that makes Corrington speculate about the whole. The "person" is the name for the phenomenological subject purged of the natural attitude just as Dasein indicates the same for Heidegger or the transcendental subject entails for Husserl. If the analogy of the phenomenological subject doubles back upon nature in Corrington's thought, then it seems that the generative-generating of nature naturing of the preformal potencies is akin to the unconscious and conscious-shaping of intentional acts as they are directed in constituting the meaning of some determination and meaning of the intended object qua nature natured.

In the eleventh proposition, Corrington claims that there is a not-yet dimension of the generative power of intentionality that is projected onto the nature-naturing and nature-natured distinction. For Corrington, what happens as "the not-yet" is but "the ultimate clearing away" for what can appear as a phenomenon. By thinking this not-yet is concealed and the thinking of nature as possessing an unconsciousness is to introduce personalistic elements back into nature about how it can change and be determined. In nature, we find ourselves. Only the openness of persons can suffice for such understanding. However, it's hard at this point not to think of nature as an agent possessing noesis and noema as well as an unconsciousness. In other words, the idealism of an unfolding process-oriented nature may emerge out of the layers of what can be. This option is fully possible and lurks because of the resources of phenomenology so employed even if we admit his contrary statements in the eleven propositions of ordinal phenomenology.

Initially, the denial of an essence tracks also with the denial of simples in nature. There's no primitively basic notion in Corrington's thought, no foundation or meta-whole that "houses and locates all of the many orders of the world."[24] In traditional Christian theology, I agree with Corrington that all ontotheologies introduce an asymmetry between God and nature, Creator and created. Ontological parity does not exist then in classical theisms. So nature becomes devalued, incapable of being seen on its own terms. In such accounts, nature is closed, determined, unchanging, reified, and static.

If the orders cocreate other ordinal complexes, then it may be possible to be guilty of what I call Alexanderian emergentism and again the appeal to phenomenology will not help. One complex of being gives rise to another complex. Just as Samuel Alexander speculated in his *Space, Deity, and Time* (1920), spacetime is the encompassing all there is, and that categorical qualities are all properties of spacetime. The categorical qualities give rise to organizations of matter. Organizations of matter give rise to life, and mind gives rise to the irreducible content to consciousness that ends in the Deity. Much like Corrington, then, and all things being equal, I could presuppose these very same Alexanderian hierarchies and build phenomenology around the possibility of my commitments being right. I can see what I presuppose in what I take to be given. What's more, the fact is we have no way of showing how hidden potencies function in Corrington, but Corrington has only claimed ontological parity about their manifestation as a way to limit what ordinal phenomenology may describe. Given the dearth of qualification about hidden potencies, the hidden potencies may not function all on the same level, but give rise to other orders of being that cannot be reduced to previous innumerable orders leading to asymmetric dependency, a laddering of Being, and ontological priority he fervently denies. A hierarchy could very much well form between these innumerable complex orders all things being equal in the same way that it did for Samuel Alexander.

In *Deep Pantheism*, one might locate this possibility in the discussion between Emerson and Plotinus. Corrington will deny Plotinus of One emerges many subsequent and layered complexity, but agreeing with Emerson. "For Emerson, nature never rests on its laurels but instead pours forth its infinite energies into the world of manifold orders."[25] What's to say that the manifold orders do not layer, and do not give rise to an ontological priority of continual emergence? As we are speculating, and not actually doing phenomenology (if I am right), we are only finding what we've speculated about, and the authority that ordinal phenomenology gives us is to simply shore up our commitments to ecstatic naturalism and deep pantheism. Such speculations are not opening up to new evidentiary possibilities.

Corrington and Emerson, then, do not give us resources to avoid metaphysical systems that give rise to ontological priority, and this might be a key reason for us to invite Corrington to respond. Unbridled fecundity might not be the right response for a world interpretation seeking to avoid any ontological foundations or final vocabularies. Perhaps another working metaphor operating a tergo behind his thought may suggest the way. And indeed, if we are inclined to see phenomenological descriptions as reports of intentional consciousness on the side of acts and the intended objects as both a constitutive duet, the original originating relationship on which my contact with the world happens, then the reluctance and criticism Corrington weighs against phenomenology might be the only possibility of what phenomenology can do. I embrace this fact.

Since phenomenological insight highlights the fact that reality is given in terms of its ontological priority, hierarchies are everywhere and structured as we encounter them. Then the goal of phenomenology is to describe this encounter as we do experience the world and the selective interests that constitute the field of attention inevitably constitute hierarchies if our experience is the basis on which we construct metaphysical systems from imaginative speculation. For these reasons, I reject the assertive interest of Corrington while also thinking that his project is not in need of any claim to phenomenology beyond thinking that intentionality is read into the nature-naturing and nature-natured distinction. What's interesting here is a possibility to get a personalism or at least proto-personalism perspective into his deep pantheism.

THE PLACE OF THE PERSONAL BECOMES THE IMPERSONAL

In this chapter, I have taken great pains to show why it is the case that phenomenology is a strategy Corrington should abandon, and that perhaps there's a proto-personalism that saturates his perspective. In this last part of the chapter, I want to completely set this aside and now urge the reader to think that the starting place of this problem is that phenomenology does not adequately feature in his perspective of nature since Corrington does not take seriously how it is that the first-person and not the third-personal form of nature governs our speculations about all there is. Instead, Corrington's metaphysics should have started, as it were, from the problem of immediate experience where also an adequate ontology can be found (and implicitly this also may be very well where he may be starting if you find like I do that there is an attenuated proto-personalism in the metaphors he uses to understand nature).

When I take a step back from the central thinkers in the pragmatist and phenomenological traditions, what we are left with is precisely the problem of immediate experience. Any philosophy that avoids the problem of immediate experience ignores the relevance of experience for a form of impersonalism that cannot be maintained.

Before outlining the problem, let me offer some brief remarks about the historical situation that conditions all forms of impersonalism on this issue. The Kantian baggage cannot be shed. As we stand in the midst of Western philosophy, there are two options, both mutually exclusive as they have been inherited from Kant. On the one hand, we can use theoretical speculation from the third-personal viewpoint—most often a form of naturalism and here what Corrington calls deep pantheism—and imagine human beings as just another object in the causal nexus of the universe. From this third-personal viewpoint, the problem of immediate experience is solved from without and outside exploring the internal contours of what it means to see human life in a personal way. The person is absorbed into the cosmos as just another configuration of the possible innumerable orders.

From the first-personal viewpoint, as Kant describes, I am aware of myself as an acting being. I can construct rational means to achieve ends. I can give myself reasons for acting. I desire. I love. I hate. I condemn. I resist. In all of these mental acts, the phenomenological traditions, the personalist tradition, and the Jamesian pragmatist offer explicit resources for the first-personal sphere to be analyzed. In conceiving myself from the first-personal sphere, the personalist acts *as if* the person is free. From within, it looks very much so. From without, it looks as if we are not free. Which one is true?

This is known as Kant's Third Antinomy of Pure Reason. For Kant, speculative reason can imagine both problems and vantage points. Moreover, the solution is nowadays almost always neuro-something, neuro-focused, privileging the uncritical acceptance of the third-personal naturalism that filters our theorizing and knowing about personal life from without while being trapped within. However, Husserl's redemptive critique of the natural attitude cannot be ignored. The fact that I think neuroscientific findings apply to my experience of X is itself a subjective accomplishment. In fact, all knowledge is an achievement from the subject who learns it, discovers it, researches it, or disconfirms it. The point is endlessly simple.

The first personal must take primacy. The fact that all knowledge is mediated by a subject who lives through it, experiences it, will experience it, and has experienced it is a transcendental feature of third-personal knowledge. Third-personal knowledge itself is, therefore, not that which engenders, but that which arises out of the constituting subject. This was Husserl's insight, and we may recover this fact when addressing the pantheistic thinking and ecstatic naturalism of Corrington. It was not that Husserl had a hard time with naturalism itself, but the uncritical attitude that all objects of knowledge can explain all aspects of experience while forgetting the constituting role played by intentional consciousness within the first person. In effect, we live within immediacy and it is only from abstracting away from immediate experience that we posit the third-personal to know and dream about nature, and I offer this is the reason that Corrington borrows from phenomenology and psychoanalysis—both of which are philosophies that mediate understanding from within, not without. While from within, we suffer from the illusion we act as if we are free. This should be reversed. While from without, we suffer from the illusion we are conditioned *as if* we are unfree, determined in relation to other causal objects.

When the personalist scholar Rufus Burrows critiqued William Stern's pantheistic personalism his objections, I think, equally apply to Corrington's perspective given the dearth of phenomenology should be denied in his account of it. Consider Burrow's words about Stern that provide an analogy I wish to offer to Corrington's thinking:

> If the pantheistic view is taken as true, sin and evil are not caused by created persons at all. Rather, they are caused by God, in whom created persons lose their sense of individuality and relative independence. Both of these—individuality and relative independence—are necessary for real freedom.[26]

Corrington is offering us a view of disenchanted nature and steering us away from the monotheistic, but also trying to preserve those elements of personal experience where sacred folds arise. In so doing, I have stayed clear of theological language even though Corrington's work is well regarded in philosophical theology. For this reason, I think Burrow's insight against Stern is very apt. If even our experiences are types of fissioning of nature naturing that emerges into complexes of persons, then I as a person become absorbed into strict metaphysical identification with the natural and deterministic processes of nature that emerge. Even still, Corrington may see that I am free in terms of having an unconscious with a hidden potency much like nature, my freedom still follows the regularity in which any Selving process can be understood. Nature's Selving and my personal selving are in this way not distinct in much the same way that sin and evil arise out of the process of that identification in the above extract. Given pantheism, evil doesn't come, then, from any complex distinction from nature, nor is there any adequate place for the phenomenology of values to occur in the very way I have explained Corrington's eleven characteristics of ordinal phenomenology. The reason why it is difficult to adequately conceptualize the place of values in Corrington's deep pantheism is that he seems to adopt a third-personal view that absorbs the real theoretical priority all philosophers and theologians should find in the personal sphere. Values come out of the personal sphere.

CONCLUSION

To conclude then, there are two overriding concerns about ordinal phenomenology present in Corrington's *Deep Pantheism* (2016). First, there is the place and function of what that phenomenology provides. Next, given the failure of ordinal phenomenology to function as Corrington thinks it does, it can be shown that his third-personal emphasis of deep pantheism does not have an adequate place for value experience at all given the dearth of phenomenology overall. This insight can only be maintained because of a transcendental commitment to the first-personal sphere. For this reason, the first problem of all philosophy is the problem of immediate experience where values are experienced conditions of all types of philosophizing. For this reason, the first-personal sphere is an inescapable perspective where all philosophizing starts. It is for this vitalist reason then that all impersonalisms should be rejected in principle.

NOTES

1. I want to thank the participants of the Twelfth Biennial Personalist Workshop for their commentary at Western Carolina University in July 2019 and the generosity of Robert Corrington's engagements with my criticisms.

2. I want to make clear that this paper is not against Robert Corrington's long time defense of ecstatic naturalism. Instead, I am only making claims in this chapter about how Corrington employs ordinal phenomenology and what he takes to be

significantly different about ordinal phenomenology than both the transcendental and existential versions in Husserl and Heidegger.

3. Robert S. Corrington, *Deep Pantheism: Toward a New Transcendentalism* (Lanham: Lexington Books, 2016), xxiii.

4. Corrington, *Deep Pantheism*, xxiii.

5. Edgar S. Brightman, *Persons and Reality* ed. Peter Bertocci (New York: Ronald Press Co., 1958), 45.

6. Corrington, *Deep Pantheism*, xxiii.

7. See Anthony Steinbock, *The Verticality of Religious Experience* (Bloomington, IN: Indiana University Press, 2009).

8. Corrington, *Deep Pantheism*, xxiii–xxiv.

9. Ibid., xix.

10. The fact that the intentional relationship can give no assurance about the reality of the world in *Ideas 1* is not a privileging of consciousness for the idealist interpretation, but on the contrary, just a sincerity about the limits of phenomenological description.

11. Robert Corrington draws heavily from Justus Buchler's pragmatist metaphysics of nature. Corrington is drawn to Buchler's principle of ontological parity and non-totalizing hierarchies. Part of Corrington's continual use of the term "phenomenology" may be more in line with this influence given Buchler's having used something like Peirce's "phaneroscopy." Either way the influence falls, *Deep Pantheism* still commits to the use of this term in the Husserlian vein.

12. Corrington, *Deep Pantheism*, xxiii.

13. Ibid., xxi.

14. Ibid., xxiii.

15. Ibid., xxii.

16. Ibid., xxiii.

17. Ibid., xxii.

18. Ibid., xxiii.

19. Ibid.

20. Ibid.

21. William James, *A Pluralistic Universe* (Frankfurt: Outlook, 2018), 4.

22. William James, *Pragmatism: A New Name for Some Old Ways of Thinking* (Cambridge, MA: Harvard University Press, 1975), 52.

23. Corrington, *Deep Pantheism*, xxiv.

24. Ibid., 4.

25. Ibid., 12.

26. Rufus Burrow, *Personalism: A Critical Introduction* (St. Louis: Chalice Press, 1999), 39.

REFERENCES

Brightman, Edgar S. *Persons and Reality*. Edited by Peter Bertocci. New York: Ronald Press Co., 1958.

Burrow, Rufus. *Personalism: A Critical Introduction*. St. Louis: Chalice Press, 1999.
Corrington, Robert S. *Deep Pantheism: Toward A New Transcendentalism*. Lanham: Lexington Books, 2016.
James, William. *A Pluralistic Universe*. Frankfurt: Outlook, 2018.
———. *Pragmatism: A New Name for Some Old Ways of Thinking*. Edited by Bruce Kucklick. Hackett Press version, 1981.
Steinbock, Anthony. *The Verticality of Religious Experience*. Bloomington: Indiana University Press, 2009.

Chapter 3

Dwelling with the Deep Ones

Lovecraftian Horror and the Selving Process

Thomas Millary

INTRODUCTION

H. P. Lovecraft's science fiction horror stories constitute an important entry in the American literary canon, an essential attempt to engage the sublime terror of nature's infinity. A psychoanalytic reading of Lovecraft can help discern the philosophical significance of his work. Robert Corrington's method of ordinal psychoanalysis provides us with a lens to understand the relationship between Lovecraft's unhappy life, his racist social perspectives, and the metaphysical insights that may be gleaned from his stories. Corrington's philosophy of ecstatic naturalism and Lovecraftian fiction both attempt to decenter the human in intriguing ways and both describe the shattering experience of recognizing nature's boundless indifference. However, in ecstatic naturalism we find a more complex and varied perspective on the powers and potencies of nature, while Lovecraft portrays them as exclusively horror-inducing. By considering Lovecraft's personal and philosophical commitments to regularity and consistency through Corrington's approach to birth trauma and the selving process, we can clarify the origins, value, and shortcomings of his imaginative fiction.

TOTALIZING NOTHINGNESS

The Shadow over Innsmouth is among the most frightening works of horror written by Howard Philips Lovecraft. The narrator of the story is a government official who carries out "a strange and secret investigation of certain conditions in the ancient Massachusetts seaport of Innsmouth."[1] The desolate coastal town has a copious amount of empty buildings and a generally foreboding atmosphere. There is something slightly off about many of its inhabitants, the narrator referring to their flat noses, narrow heads, and bulgy eyes as the "Innsmouth look."[2] He hears a disturbing tale of the Deep Ones, an ancient amphibious race who can reproduce with humans, spawning individuals who are born human but eventually morph into those aquatic immortal beings. In a harrowing sequence, the protagonist flees the town for his life in the middle of the night, narrowly escaping unharmed. However, over the years, first through disturbing dreams and then through physical changes, he comes to learn the awful truth: that his own ancestry is mixed with these creatures and he himself is transforming. His terror fades to an unnerving expression of joy in the final words of the story. Referring to a relative who shares his condition, he declares,

> Together we shall go to marvel-shadowed Innsmouth. We shall swim out to that brooding reef in the sea and dive down through black abysses to Cyclopean and many-columned Y'ha-nthlei, and in that lair of the Deep Ones we shall dwell amidst wonder and glory forever.[3]

H. P. Lovecraft's life was brief and dark, haunted by intense anxieties and vividly nightmarish dreams. He spent the majority of his life in Providence Rhode Island, where he died of cancer at the age of forty-six in 1937. He was a prolific writer of horror stories, creating works that (while met with modest success within his lifetime) have significantly influenced the genre. Stephen King has praised him as the "twentieth century's greatest practitioner of the classic horror tale,"[4] referencing "his shadow, so long and gaunt, and his eyes, so dark and puritanical, which overlie almost all of the important horror fiction that has come since."[5] Lovecraft's life was troubled, yet he managed to channel his personal darkness into the creation of what is perhaps the second most iconic body of works of American literary horror, after those of his great influence, Edgar Allan Poe.

Lovecraftian horror is rooted in a particular conception of nature. Frequently bringing science fiction elements into his stories, Lovecraft was deeply invested in the project of modern naturalistic science, understood as rapidly expanding humanity's ability to understand the cosmos. He quickly came to the conclusion that the world revealed by the natural sciences was one totally devoid of intrinsic meaning or spiritual purpose, once remarking, "By my thirteenth birthday I was thoroughly impressed by man's impermanence and insignificance."[6] That conviction remained with him throughout his life, a worldview known as *cosmicism*, which "relegate[s] the entire history of the human race to an inessential nanosecond in the realm of infinite space and time."[7] Amazed by the boundlessness of the natural world, he believed the only decent option for humanity is for us to become content with our ultimate lack of meaning within it.

However, contentment was never something he ever achieved, nor is it something available to the ill-fated protagonists of his creepy and richly atmospheric stories. An often-quoted passage is an opening paragraph to his most well-known story, "The Call of Cthulhu":

> The most merciful thing in the world, I think, is the inability of the human mind to correlate all its contents. We live on a placid island of ignorance in the midst of black seas of infinity, and it was not meant that we should voyage far. The sciences, each straining in its own direction, have hitherto harmed us little; but some day the piecing together of dissociated knowledge will open up such terrifying vistas of reality, and of our frightful position therein, that we shall either go mad from the revelation or flee from the deadly light into the peace and safety of a new dark age.[8]

Lovecraft was a committed atheist, regarding religion and spirituality as superstitions, or archaic beliefs that have no value in a scientifically aware age. At the same time, he was incredibly pessimistic about the prospects of humanity being able to adjust to an evaluative paradigm of cosmicism. He found it much likelier that being disabused of our myths concerning our own cosmic importance would drive us collectively to insanity, as reflected by his many stories where characters go mad after glimpsing too much truth about the universe. Robert Corrington describes Lovecraft's work as "a world in which horror gets the last word."[9]

The description of nature expressed in Corrington's philosophy of ecstatic naturalism is expansive in a manner that radically decenters the human: "We are always *in medias res* with the origin or the goal forever shrouded in thick mist by the very nature of things."[10] As such, the scope and trajectory of human experience (what Corrington terms "the human process") is ordinally located within the boundless series of complexes of which nature is comprised.

In *Nature and Nothingness*, Corrington discusses how Lovecraftian horror is a powerful artistic expression of the terrifying effect that nature can have upon the human process. One of the four variations on nothingness that Corrington examines is totalizing nothingness, which denotes the terror of human assumptions about nature being shattered. To be exposed to totalizing nothingness is "to encounter the total eclipse of the human process and to sink into the permeating sense of nature's total indifference and primal uncaring aspect."[11] Corrington explores totalizing nothingness through a careful reading of stories from both Lovecraft and Poe. He notes an important contrast between the psychoanalytic Poe and the more cosmically minded Lovecraft, in that "Poe probes into the totalizing nothingness within, whereas Lovecraft devotes his literary energy to manifesting the totalizing nothingness without."[12]

LOVECRAFT'S DARK DREAMS

Lovecraft is relevant to naturalist philosophy because of the vividness with which he portrays the nothingness that humans so often encounter in nature. Frameworks such as Corrington's ecstatic naturalism are unflinchingly Darwinian in perspective, making no attempt to minimize the ubiquity of suffering and violence in evolution. Corrington reminds us that a crucial implication of Darwinism is that "there is no guarantee that the current human species will escape extinction."[13] Lovecraft himself was an admirer of Darwin, gaining much inspiration for his cosmicism from Darwinian denial of anthropocentric teleology. In light of nature's more tragic qualities, that philosophy can be quite persuasive. In Corrington's view, "horror lies at the basis of evolution and human interaction, from warfare, to genocide, to terrorism."[14] Lovecraftian mythology is filled with unsettling metaphors for how it feels to be fully exposed to such traumatic realities.

Still, the enthrallingly imaginative aspects of Lovecraft's work point toward a compelling tension within him. Not only was Lovecraft a Darwinian atheist but he passionately espoused a decidedly mechanistic view of nature. The attributes of nature that most impressed him were its infinity and its eternity, the sheer scope of the cosmos and its high degree of determinism. So committed was he to the notion of nature's regularity that even the theories of Einstein made him severely uncomfortable. Yet, the imagery and concepts he employed within his stories to represent his cosmicism are dynamic and creative. The entities that populate his Cthulhu mythos may be metaphors for cosmic indifference but their descriptions often feel overwhelmingly ecstatic, even as they represent a static universe.

In "The Call of Cthulhu," the narrator uncovers the horrible truth of Cthulhu, the giant tentacled entity who lies dreaming in the sunken city of R'lyeh. Cthulhu is one of the Great Old Ones, extraterrestrials who populated the earth long before humankind. The city of R'lyeh is characterized by incomprehensible non-Euclidean geometry and Cthulhu's nightmarish presence is enough to shatter the human mind. By the tale's conclusion, the narrator is expecting an unpleasant end for humanity. "Who knows the end? What has risen may sink, and what has sunk may rise. Loathsomeness waits and dreams in the deep, and decay spreads over the tottering cities of men."[15]

While the tentacled likeness of Cthulhu has proven to be the Lovecraftian invention with the most staying power in mainstream pop-culture, nature as portrayed in these stories is home to myriad alien creatures, powers, and forces. The Great Old Ones pale in comparison to the Outer Gods, who are as far beyond Cthulhu as Cthulhu is beyond humanity. These include Nyarlathotep ("the crawling chaos),[16] Yog-Sothoth (who "knows the gate . . . is the gate . . . [and] is the key and guardian of the gate"),[17] and Azathoth ("the last amorphous blight of nethermost confusion that blasphemes and bubbles at the center of all infinity").[18]

These are intriguingly imaginative expressions of a supposedly meaningless universe. I have no doubts as to the sincerity of Lovecraft's dedication to a mechanistic naturalism but it is evident that there was much at work in his unconscious. As mentioned, he was a vivid dreamer throughout his life, going through a period as a young boy where he constantly experienced nightmares where he was tormented by malevolent abstract entities.[19] That intense dream imagery stayed with him through his whole life, occasionally providing inspiration for his fiction. His waking life was similarly uneasy, as he was plagued by frail health, low self-esteem, and financial difficulties, all before a painful death at a young age.

Furthermore, he was aggressively racist, even by the standards of his time. As he was wary of shifts within scientific paradigms, he was even more fearful of shifts within society. Suspicious of virtually everyone who was not a white New Englander, Lovecraft was horrified by immigration and the other cultural transitions he witnessed during his lifetime, which he saw as threats to the classical aristocratic Anglo-Saxon culture he was so fond of. These tendencies are often reflected in his writings, whether through the portrayal of denigrating stereotypes or in projections of these anxieties unto the monsters he created. The crossbreeding between humans and the Deep Ones in *Innsmouth* is an obvious metaphor for his fear of the white gene pool being "polluted" by interracial marriage. As a biracial person myself, this aspect of Lovecraft's story is particularly interesting to me. Even as it horrified him, Lovecraft was more sensitive than many to the conceptual resonance between racial hybridity and capacity of humanity to be transformed by nonhuman forces of nature.

Lovecraft's life was permeated by alienation. His inability to affirm himself spawned a mistrust of other human beings and an outright loathing of those noticeably different from him. Comforted only by the familiar, it is no wonder that he clung to a static ideal of both culture and the cosmos. Knowing this does much to explain why he experienced "a vision of nature which makes it horrible *in itself*" (emphasis in orginal).[20] Although Lovecraft's depictions of totalizing nothingness are imaginative reminders of the tragic aspects of nature, his inability to experience the orders of nature and nature's unconscious as anything but traumatizing was likely the result of his particular psychological complexes.

Corrington's term for "the overall arc and trajectory of the human process"[21] is "selving," something that Lovecraft was almost entirely unable to undergo. In Corrington's account of the self, we are haunted by the birth trauma, the shock of being ejected from "the oceanic bliss of the womb."[22] For selving to take place, one must be willing to move "from the state of dreaming innocence in the literal and symbolic womb . . . to the fullness of life amidst the various potencies of *nature naturing* (emphasis in original)."[23] Lovecraft was more haunted by birth trauma than most. His longings for familiarity and regularity were manifestations of the desire to impossibly return to a lost condition of dreaming innocence. His reactionary social views, his pessimistic view of nature, and his brilliantly conceived fiction thus share a common origin.

In tales such as *Cthulhu* and *Innsmouth*, Lovecraft portrayed the depths of the sea as the site from which otherworldly terrors emerge. That dark association is understandable, considering his antagonistic relationship with unconsciousness of all kinds, including the unconscious of nature itself. His sensitivity to the sovereignty of nature over the human experience makes Lovecraftian horror a helpful corrective to overly honorific or romantic accounts of nature, which are unable to speak honestly of totalizing nothingness or the true otherness of the unconscious. Even so, Corrington's method of ordinal psychoanalysis can help to reveal an even fuller portrait of the world.

WONDER AND GLORY

Corrington writes that "Ordinal Psychoanalysis is open to the innumerable ways in which the selving process transacts with the deepest pulsations of *nature naturing*" (emphasis in original).[24] Those innumerable ways include the sublime terror of totalizing nothingness but also interactions with nature that can provide humanity with a sense of direction and guidance. Lovecraft never managed to undergo the complete spiritual journey of selving but it is possible for us to do what he did not and seek "integration of the potencies of nature and the archetypes of the collective unconscious, along with the complexes of the personal unconscious."[25] Horror may be a true and undeniable component of the human process but it does not exhaust the selving experience.

Here we begin to diverge from Lovecraft's cosmicist paradigm of a world utterly bereft of cognizable meaning. If probing the deep waters of nature's beauties and terrors can be a source of guidance, to what are we guided? While ecstatic naturalism takes great care to stay within the limits of phenomenological evidence in describing the aspects of nature that exhibit purpose, Corrington does advance the concept of "the Wisdom." The Wisdom is a chastened form of the Emersonian Oversoul, characterized as "wise beyond our understanding even if 'itself' is on its way to more wisdom . . . a deep well of insight that is passed on to the human process."[26]

It seems that more might be at work within selving than just our individual paths from unconsciousness to greater self-awareness. In undergoing selving, we contribute to the evolution of the Wisdom, an impetus toward maturation, purpose, and even healing. Though we experience this presence as a guide to human development, it is a reality that stretches far beyond the human. As a vehicle for selving, the Wisdom necessarily derives its efficacy from its ability to provide humanity with lessons from the nonhuman orders of nature. Our ability to receive such insight and transmit the experience of our unique species journey into nature gives humanity more cosmic significance than seen by Lovecraft. However, there remains an inescapably unsettling aspect to this reassurance. As a connecting force between humanity as it exists and our future integrations with the powers and potencies of nature (even those that we experience as horrifying), human fate might include a reckoning with the shattering forces feared by Lovecraft. We cannot rule out the possibility that the Wisdom has tentacles.

Even the unhappy life of H. P. Lovecraft seems to have contributed to the human capacity for selving. Despite his undeniable racism and general psychological unhealth, his imagination produced a plethora of creative representations of nature's ecstatic powers. As dismissive as he was of spirituality, he envisioned a cosmos in which a multiplicity of forces exist that humans can encounter and be transformed by. We can revel in the dark enchantment of his works of horror while recognizing the incompleteness of his worldview. The sublimely terrifying forces of nature are not always wholly indifferent but sometimes empowering and healing. Likewise, dynamic shifts within culture that generate increased hybridity can be cause for joy rather than anxiety.

The ending of *The Shadow over Innsmouth* represents Lovecraft's ultimate fear of being exposed to the unknown and unfamiliar. The story's narrator experiences the worst outcome Lovecraft could conceive of, becoming simultaneously internally intertwined both with humans different from him and the dark forces of nature. Yet as a kind of Lovecraftian optimist, I feel unironically sympathetic to the narrator's concluding mindset. He confidently embraces the strangeness of his heritage and plans to journey to the lair of the Deep Ones and dwell with them amid wonder and glory forever. Does that fate really sound so bad?

NOTES

1. Howard Phillips Lovecraft, "The Shadow over Innsmouth," in *The New Annotated H. P. Lovecraft*, ed. Leslie S. Klinger (New York: Liveright Publishing Corporation, 2014), 573.
2. Ibid., 590.
3. Ibid., 642.
4. Curt Wohleber, "The Man Who Can Scare Stephen King," in *American Heritage* 46, no. 8. December 1995. http://www.americanheritage.com/content/man-who-can-scare-stephen-king.

5. Ibid.
6. S. T. Joshi. *I Am Providence: The Life and Times of H. P. Lovecraft* (New York: Hippocampus Press, 2013), 182.
7. Ibid., 459.
8. Howard Phillips Lovecraft. "The Call of Cthulhu," in *The New Annotated H. P. Lovecraft*, ed. Leslie S. Klinger (New York: Liveright Publishing Corporation, 2014), 124.
9. Robert S. Corrington. *Nature and Nothingness: An Essay in Ordinal Phenomenology* (Lanham, MD: Lexington Books, 2017), 48.
10. Robert S. Corrington. *Nature's Sublime: An Essay in Aesthetic Naturalism* (Lanham, MD: Lexington Books, 2013), 80.
11. Ibid., 27.
12. Ibid., xi.
13. Ibid., 5.
14. Ibid., 54.
15. Lovecraft. "Call of Cthulhu," 157.
16. Howard Phillips Lovecraft. "Dream-Quest of Unknown Kadath," in *The New Annotated H. P. Lovecraft: Beyond Arkham*, ed. Leslie S. Klinger (New York: Liveright Publishing Corporation, 2019), 331.
17. Howard Phillips Lovecraft. "The Dunwhich Horror," in *The New Annotated H. P. Lovecraft*, ed. Leslie S. Klinger (New York: Liveright Publishing Corporation, 2014), 358–359.
18. Lovecraft. "Dream-Quest," 332.
19. Joshi. *I Am Providence*, 58.
20. Corrington. *Nature and Nothingness*, 43.
21. Corrington. *Nature's Sublime*, 32.
22. Ibid., 51.
23. Ibid., 32.
24. Ibid., 9.
25. Ibid., 47.
26. Robert S. Corrington. *Deep Pantheism: Toward a New Transcendentalism* (Lanham MD: Lexington Books, 2016), 87.

REFERENCES

Corrington, Robert S. *Deep Pantheism: Toward a New Transcendentalism*. Lanham, MD: Lexington Books, 2016.

———. *Nature's Sublime: An Essay in Aesthetic Naturalism*. Lanham, MD: Lexington Books, 2013.

———. *Nature and Nothingness: An Essay in Ordinal Phenomenology*. Lanham, MD: Lexington Books, 2017.

Joshi, S. T. *I Am Providence: The Life and Times of H. P. Lovecraft*. New York: Hippocampus Press, 2013.

Lovecraft, Howard Phillips. "The Call of Cthulhu." In *The New Annotated H. P. Lovecraft*, edited by Leslie S. Klinger. New York: Liveright Publishing Corporation, 2014.

———. "The Dream-Quest of Unknown Kadath." In *The New Annotated H. P. Lovecraft: Beyond Arkham*, edited by Leslie S. Klinger. New York: Liveright Publishing, 2019

———. "The Dunwhich Horror." In *The New Annotated H. P. Lovecraft*, edited by Leslie S. Klinger. New York: Liveright Publishing Corporation, 2014.

———. "The Shadow Over Innsmouth." In *The New Annotated H. P. Lovecraft*, edited by Leslie S. Klinger. New York: Liveright Publishing Corporation, 2014.

Wohleber, Curt. "The Man Who Can Scare Stephen King." *American Heritage* 46, no. 8 (December 1995). http://www.americanheritage.com/content/man-who-can-scare-stephen-king.

Part II

FACING SUFFERING AND VIOLENCE

ECSTATIC DIFFERENCE AND EDUCATIONAL HEALING

Chapter 4

On Being Sunk?

Desmond Coleman

INTRODUCTION

In this chapter, "On Being Sunk?," I engage W. E. B. Du Bois's famous ventriloquizing of the question perpetually thrown at him by the progressive whites of his time: "How does it feel to be a problem?" Instead of approaching that question through the metaphor of the Veil, as he does, however, I approach and explore this question through the metaphors of "sunkenness" and "the Sunken Place." The former, "sunkenness," is a concept-metaphor used by G. W. F. Hegel throughout his work to speak of the determinate "mental and spiritual characteristics" of the "Negro race." The latter, "the Sunken Place," is a concept-metaphor used by writer-director Jordan Peele in his blockbuster *Get Out*. While Hegel and Peele are writing centuries apart, and while it is unlikely that Peele borrowed this concept-metaphor from Hegel, the resonances and dissonances of their use of the notion of sunkenness vis-à-vis the meaning of blackness raises important questions about the logical and historical conditions of possibility of this connection (between "blackness" and "sunkenness"), and about the relationship between this connection and the lives of those people we call "black."

ON BEING A PROBLEM

I would like to begin my chapter by pointing out the obvious thing that race, class, and gender are issues our society must address. And, that they are problems that should be addressed, even if not solved or resolved, in politically correct scholarly discourse. It is therefore safe to say that you as reader are expecting that I will address at least one of these problems. It is also safe to say, I think, that given my appearance, you (and the "you" here is more structural than anything) are expecting that I will, at *very least*, address these problems as they relate to my race as a black person. If this is the case, it is as if I am being asked a question that W. E. B Du Bois was also asked, with equal indirectness.

Indeed, these words, published in 1903 in *The Souls of Black Folk*, are strangely fitting for this occasion. This is how Du Bois articulated the problem of the so-called Negro Problem:

> Between me and the other world there is ever an unasked question: unasked by some through feelings of delicacy; by others through the difficulty of rightly framing it. All, nevertheless, flutter round it. They approach me in a half-hesitant sort of way, eye me curiously or compassionately, and then, instead of saying directly, How does it feel to be a problem? they say, I know an excellent colored man in my town; or, I fought at Mechanicsville; or, Do not these Southern outrages make your blood boil? At these I smile, or am interested, or reduce the boiling to a simmer, as the occasion may require. [But to] the *real* question, *How does it feel to be a problem?* (emphasis mine). I answer seldom a word.[1]

Luckily for us, or at least for *me*, Du Bois eventually decided to answer the question perpetually thrown at him by the progressive whites of his time, and more violently by the nonprogressive population who so ardently put faith in their "Whiteness": How does it feel to be a problem? Or, more specifically, "How does it feel to be the Negro (i.e., Black) problem?"

Taking Du Bois as my inspiration, I will address what was called the Negro Problem, or, the problem of blackness, not by means of the metaphor of the Veil, as he does, but by means of, as the title of my presentation suggests, the metaphors of *sunkenness* and the *Sunken Place*.

THE SUNKEN PLACE AS CONCEPTUAL METAPHOR

I first stumbled upon the concept-metaphor of "sunkenness" in G. W. F. Hegel's *Philosophy of Mind* (published in 1816). Then, again, in his *Elements of the Philosophy of Right* (completed in 1820). I thought it was interesting, especially given to what and/or whom he was applying this concept-metaphor, but I didn't pay it too much mind, or understand it to be anything really worth thinking about—until I saw Jordan Peele's 2017 comedy-horror film *Get Out*, where I encountered a version of it again, this time as "the Sunken Place."

Get Out, for those who may be unfamiliar with it, has the honor of being the highest-grossing movie debut for an original screenplay in the history of the Oscars. It grossed nearly *three hundred million* dollars across the globe, and *primarily*, I speculate, because of the potency of this "Sunken Place" metaphor.

But what is it, the Sunken Place? And what does it have to do with Hegel? And what do either of these (the Sunken Place and Hegel) have to do with the so-called Negro Problem, the problem of "blackness," and what Du Bois described as the "strange feeling" of *being* that problem? To answer *these* questions, I find it necessary to give a brief synopsis of what happens in the movie up to the pivotal Sunken Place scene. In the film, Chris Washington, an aspiring professional photographer, and the black male protagonist of the film, has been dating Rose Armitage, a white woman, for approximately five months. But, he has not yet met her parents. In an early scene of the movie, Chris expresses that he is nervous about meeting them for the first time, and asks Rose if she has told them that he is black.

Rose admits that she has not but attempts to comfort him with the fact that her dad, a surgeon, voted for Obama *twice* and that he would've voted for him a *third* time if that were possible. In other words, Chris has nothing to worry about. Her father's house is a "safe space."

When they finally arrive at the house, a humongous estate, Rose's father, Dean, greets Chris at the door with a "My man!"[2] Rose's mother, Missy Armitage, on the other hand, appears to be more levelheaded, less concerned to affirm Chris's blackness, and therefore also coming as more authentic. Her character stays in the background until about twenty minutes into the movie, in a scene where Chris, Rose, Dean (Rose's father), and Missy (her mother) are drinking tea outside in their scenic backyard.

In this scene, Rose's parents ask Chris about *his* parents. Chris tells them that his dad "wasn't really in the picture" and that his mom was killed in a hit-and-run accident when he was eleven. Rose, knowing how traumatic this is for Chris, grabs his hand in consolation.[3] Her parents, playing the role of emphatic fellow human beings, give their condolences. Rose's mom assures him, as she taps on the glass of tea in front of her, that "it's okay. We don't have to talk about that."[4]

But Chris is already anxious. He begins to tap and scratch on the table. Rose's dad recognizes this as the nervous habit of a smoker, and asks Chris if he smokes. Chris admits that he does but says he is trying to quit.

At this point we learn that Rose's mom is a psychiatrist, and that she has developed a method of hypnosis through which she can help Chris kick his habit. Rose's dad admits that he himself used to smoke and that it helped him. Rose, again, the conscious girlfriend, recognizes the awkwardness of this suggestion, and tells her dad that perhaps Chris doesn't want strangers messing around in his head. They change the conversation and begin talking about "the big get together" that is happening the next day. This get together, as Dean explains it, is a continuance of a family tradition started by his deceased father and mother, Rose's grandparents, where they invite friends over for an all-day celebration.

In the middle of this conversation, a black woman, Georgina, acting as the housekeeper, walks up smiling, happy as can be, and begins refilling everyone's glasses with tea. When she gets to Chris, however, something changes. She appears to get lost in her own thoughts as she pours, a reverie of sorts, and accidently overfills his cup. Realizing this, she snaps out of it and becomes extremely apologetic and upset. Rose's mother instructs Georgina to "go lay down and just get some rest."[5] Georgina, back to her happy self, responds in a rather servile manner: "Why yes, I think I will."[6] It is at this moment that Chris begins to suspect that something strange is going on at the Armitage house.

I think this is enough background information to give you a sense of what's going on in this scene, the "Sunken Place" scene.

In this scene, Missy Armitage puts Chris in what she calls the "Sunken Place." As is clear, the Sunken Place is a place "inside" of Chris's being, indeed, inside of his body, but also, note, "in the floor." It is a place in which, as the scene makes vivid, Chris loses control of his body, and loses the capacity to express himself in any way.

Later in the film, we discover that the Armitage family has been luring black people to the Armitage house, kidnapping them, and putting them in the Sunken Place, all in order to auction their bodies to white people, and a few Asians, so that they can use whatever "natural gifts" the black body being auctioned possesses. We learn that this "service" was created by Rose's grandparents and that the bodies of Georgina, the black housemaid, and Walter, the black yard keeper, are both inhabited by Rose's grandparents. We also learn that the "big get together" is really an auction, and that Chris is the black body being auctioned this year. Around fifty or so white people (and one Asian man!), of all ages, genders, abilities, and classes, meet and greet Chris, each subtly communicating what it is they like about his body.[7]

Eventually, when Rose and Chris are away from the party, the auction takes place. The winner of Chris's body is a blind art dealer, who confesses that unlike the others, *he* is not racist. He just wants "those things" Chris "sees through." His *eyes*.

But, Chris escapes. He eventually "gets out" as the title of the film indicates. But before talking about that, I want to address a question which you may be already asking yourself: What does any of this have to do with Hegel?

HEGELIAN *VERSUNKEN* AND ALCHEMICAL BLACKNESS

Well, let's read from the two texts mentioned just a while ago. First, from the *Philosophy of Mind*, published in 1817. There, in the first section titled

"Anthropology," in the midst of a discussion of the mental and spiritual characteristics of the races, Hegel writes that Africans "do not attain to the feeling of human personality, their mentality is quite dormant, remaining *sunken* (German: *versunken*) *within itself* (emphasis mine)."[8] He goes on to say, in the second part of the sentence, that this sunkenness of the African mind is in immediate correspondence with the "compact undifferentiated mass of the African land." Now, though he does not say that African land is *versunken*, the logic of sunkenness is still at work. He says that African land is a "compact undifferentiated mass."[9] It is possible, perhaps even necessary, to trace a relation across disciplinary and spatiotemporal boundaries here, and say that Hegel is positing African land as a kind of *black hole*, insofar as a black hole is nothing other than the sinking-into-itself—of a sufficiently compact mass. Within this imaginary, African land is a black hole on earth.

But this is "only" what he says about Africa and "Africans as such": their mentality is sunken within itself, and that this sunken-within-itself-ness corresponds immediately with the sunken-within-itself-ness of the African land. He also talks about "the will of the slave" in another text, in *Elements of the Philosophy of Right*. But before reading what he says about "the will of the slave," I want to highlight the role that will play for Hegel in the *Origin of Right*, or, of what we would call social (or ethical-political) *justice*. Early in the Introduction he writes:

> The basis of right (that is, of social justice) is the *realm of spirit* in general and its precise location and point of departure is the will; the will is free, so the system of right (of systemic social justice) is the realm of actualized freedom (emphasis mine).[10]

The German word translated as "basis," here in "the basis of right is the realm of spirit, its precise location and point of departure is will," is *Boden*, which could also be translated as soil, ground, element, and even land. It is in *this* context that Hegel talks about the will of the slave, or more accurately, the will of the *African* slave. So what does he say?

Hegel says that the will of the slave is "*sunk* into its condition" (emphasis mine).[11] But what exactly is this slave's condition? First, because this slave is African, its condition is already one of sunkenness. It's mentality, that is, the way "it" thinks, is sunk within itself. And, as we have already seen, Hegel imagines the sinking-into-itself of the African mind to be an immediate result of the sinking-into-itself of the African land. But *this* African is *also* a slave, removed, but not emergent, from its sunken homeland. Not "emergent" because this transatlantic movement of the African slave from its sunken homeland is not from this point of view an *un*-sinking. It is not a movement of progress. If anything, according the logic of sunkenness violently brought to bear on the African slave in the *Philosophy of Right*, this transatlantic movement is a *recession*. Indeed, the word used in reference to the African slave is not *versunken* but *versenkte*, which could also be translated as "recessed." The African slave's will is recessed into a condition of *absolute* uncanniness, *unheimlich* (literally *un-home-like-ness*), haunted not only by the will of the master but also by its own a priori sunkenness as an African.

It is necessary now to take a quick and short detour and follow the logic of sunkenness to Hegel's *Logic*, where he talks more explicitly about the ontology of conditionality brought to bear on the African slave in the above passage.

In the second subdivision of the *Encyclopedia Logic* titled "Essence as the Ground of Existence," Hegel in so many words identifies the logic of contingency and conditionality with the logic of sunkenness. He writes,

> When we speak of the *condition* of this or that matter, this has double implication: first, something-there, and existent, or in general something immediate; and secondly, the *destination* of this immediate being is to be sublated and to *serve* for the realization of another one (emphasis mine).[12]

He continues a few sentences later: "The conditions that *sacrifice themselves*, *go under* and are *used up*, only to come together with themselves in the other actuality" (emphasis mine).[13] And again, "As conditions, the immediate circumstances *go under* (emphasis mine), but at the same time they are *also preserved as the content of the matter*."[14]

I will return to this idea of the preservation of that which has gone under as the content of the matter later. But for now, I think we can intuit here an intimacy between the logic of sunkenness and the logic of conditionality. We see that Hegel interprets the condition(s) of an actuality, or of an event, to be sunk.

It might be harder, however, to fully appreciate the intimacy between the logic of sunkenness and contingent conditionality and what Hegel says about "the African," the African slave, and "the Negro." To help us here, it is necessary to look at another passage from the *Encyclopedia Logic*. In a section devoted to delimiting the limits of Immanuel Kant's subjective idealism, Hegel calls Kant's notion of "the thing in itself" (*das ding an sich*) a caput mortuum. He writes that the thing-in-itself, as a caput mortuum, is a "byproduct of thought, such as accrues when thought is carried on to abstraction unalloyed."[15] The meaning of caput mortuum and the resemblance between the meaning of caput mortuum and the logic of sunkenness expressed elsewhere Hegel's text and corpus is crucial here.

Caput mortuum is Latin and would be translated in English as deadhead. It was used by chemists and alchemists to refer to the precipitate left at the bottom of a retort after a process such as sublimation. Within "spiritually" oriented alchemy, caput mortuum, or deadhead, was also called the *nigredo nigrius nigro*, the blackness blacker than black. While, on the one hand, Hegel criticizes Kant's notion of the thing-in-itself, on the other hand, Hegel reestablishes the process by which the thing-in-itself is posited as a metaphysical process and not merely a subjectivity-cognitive process. In other words, the thing-in-itself occupies the position of the *substance* produced through the emergence of *subjectivity*; the substance that "goes under" and is "used up," that is sacrificed, sublated, and suspended (*Aufhebung*), by subjectivity.

Thus, within this alchemically informed interpretation of Hegel's metaphysics there is at least conceptual resonance between the thing-in-itself, substance, blackness (caput mortuum as *nigredo nigrius nigro*), and sunkenness. What is more, this resonance also establishes a homology between these terms and the *versunken* Africa, "Negro," "African," and slave in Hegel's social and political writings. "The Negro" is the thing-in-itself, the substance, the deadhead, the blackness blacker than black of social and political progress. Thus, the "Sunken Place," according to this interpretation of Hegel, is "blackness" or "negro-ness" itself.[16]

This leads us to raise the question, as you may have already anticipated, of whether this conception of the Sunken Place as blackness resonates with, or even provides philosophical context for, Jordan Peele's conception of the Sunken Place? Luckily, we have the transcript of an interview where Peele indicates how he understands this metaphor. The Sunken Place, he says, is a "metaphor for *the system that is suppressing the freedom of black people*" (emphasis mine).[17] If we follow the logic of the metaphor as interpreted here, then the Sunken Place is a concept-metaphor for what we call "*anti*-blackness," and *not* blackness itself.

We arrive, then, at a clear difference, even a *contradictory* difference between the Sunken Place as *anti*-blackness in Peele's *Get Out* and Hegel's *versunken* as *blackness* itself. Because of this, I am tempted, as I think most people are or would be, to side with Peele, and simply identify the Sunken Place with anti-blackness. But things are not so simple and one-sided. The concept of "anti-blackness," which suggests that blackness exists prior to it, requires us to ask, "What is this 'blackness' that anti-blackness is against?" It is not so easy to escape the idea that what "anti-blackness" is antagonistic toward is sunkenness, and sunkenness precisely as Hegel understands it: as conditionality, as contingency, as devaluation, as going under, as death—in short, as "the entirety of the negative side of Hegelian dialectic."

Remember, the primary drive and motivation of the Armitage family, and of those for whom they are providing their "service," is escaping everything involved with contingency and conditionality. Rose's grandparents inhabit the bodies of two black domestic workers. They didn't/don't want to die; they wanted immortality. One young white woman wanted Chris's body—really, his penis—for her husband, who, though rich, was old, and apparently "impotent." Or perhaps is better to say she wanted Chris's penis for herself. She and her husband both wanted to experience an erection, another notorious contingency. The art dealer who finally won Chris's body was blind. He wanted to *see*, yet another biological contingency. All of these—death, erectile dysfunction, blindness—are, for these people, dimensions of lived sunkenness. That is, they are dimensions of *blackness* if we trace blackness through Hegel to alchemical nigredo, as I think is necessary to achieve an adequate historical understanding of the present.

But even in light of this, it is not necessary to just side with Hegel, and simply posit the Sunken Place as blackness. Instead, it seems to me that the logic of blackness and anti-blackness is dialectical; that the "Sunken Place" just *is* this dialectic of blackness and anti-blackness. One is "in" the Sunken Place insofar as one's *mentality*, that is, the mode of one's *mind*, one's *body*, one's *flesh*, one's *life*, or one's *being* is reduced to this dialectic and the contingent labor that it demands.

This is, of course, still a "Hegelian" reading of the Sunken Place. When Hegel says that the ontologically enabling conditions of any event "go under" and are "used up," but are also "preserved as the content of the matter" (*aufgehoben*), he is saying, in other words, that blackness is that which "goes under" and is "used up," but is also *preserved* as the content (or even the *heart*) of the matter, or of the event, which we would call anti-blackness—that is, blackness as sunken becomes the driving force of a progressive dialectic.

So how does one *get out* of this bind? How does one *get out* of this dialectic of blackness and anti-blackness, and escape the pull of the Sunken Place? Is it possible?

WEAPONIZING DISCOURSE OR NATURAL GRACE: GETTING OUT AS AN ECSTATICALLY INDIRECT NARRATIVE PRACTICE OF FREEDOM

We can look again to the film, and to what Peele says about the film in the interview I mentioned earlier. The title of that interview is especially relevant: "Seven Director's 'Weaponize Storytelling' to Stand against Hatred and Violence." The scare quotes within which I placed "Weaponize Storytelling" are not mine. Nor even the author of the article. They come from Peele himself. "Story," Peele says, "is one of, if not the most important tool, [or] *weapon*, we have against hatred and violence" (emphasis mine).[18] Kathryn Bigelow, director of the film *Detroit*, also part of the interview, responds to Peele here, affirming his sentiment, saying, "Especially now. You're almost weaponizing storytelling in order somehow to contextualize the unthinkable." Peele responds saying, "And story promotes empathy, right?"[19]

A straightforward interpretation Peele's statement would suggest that *Get Out* is supposed, at the very least, to be a story that garners the empathy of the "anti-black." But this interpretation has an uncanny resemblance to what takes place in the Sunken Place scene. Remember what happens there? Missy, Rose's mom, a psychiatrist, presents herself as an empathetic listener, as a safe space of conciliation and healing for Chris's trauma. But what happens? She sinks him: under the pretense of listening, and of allowing Chris to tell his story of trauma. She sinks him. In fact, she sinks him *by means of* his story.

In this way, the film instead appears to be a warning against this very assumption that telling stories of black trauma is "the most important tool, [or] *weapon* (emphasis mine), against hatred and violence," against, in this case, the pull of the Sunken Place.

Is there *another* way out? One that escapes the dialectic of blackness and anti-blackness, and the labor it demands of the minds, bodies, and lives of we who are called black?

We can turn again to the film. Chris, as I mentioned, *does* eventually get out. But how? He realizes, right after he learns that the blind man has won his body, that the cotton from the arm of the Sunken Place chair is poking out. He then uses this cotton to plug his ears, preventing the scratching of the spoon against the teacup from reaching his ears. But why was it poking out of the chair? Remember the Sunken Place scene. As Chris begins narrating his trauma to Missy Armitage, he also begins scratching the chair.

We get a quick flashback to the traumatic event, the night his mother died. On that night too he scratched the *wooden* chair in which he was sitting. Apparently, this is a nervous habit that emerged as a result of the traumatic experience of his mother's death, a nervous habit which was eventually sublimated into his smoking habit. As a nervous habit, this scratching was and *is*, in fact, Chris's attempt to resist sinking: into a depression, into melancholia, into a sunken place. But the repetition of this nervous habit, this struggle and/or resistance at the edge of Sunkenness and the Sunken Place, *this time*, resulted in the tearing of the Sunken Place chair, the revelation of cotton, and the possibility of an escape.

It is important to note, though, that the tearing of the Sunken Place chair and the revelation of the cotton falls outside of Chris's intentionality. He did not intend to tear the chair. This was the unintended consequence of the repetition of a nervous habit in a new context, under, I must say, new "material conditions": the Sunken Place chair in which he was sitting was tearable.

But how should we interpret this? This unintended, unforeseen, unpredictable revelation of a way out? Should we interpret it as luck? As chance?

Perhaps. But, I prefer a Corringtonian formulation. This revelation of a way out is the effect of a potent intertwining, or dare I say a "sacred folding," of what Corrington calls "natural grace" and the "grace of the spirits."[20] This revelation was "natural" insofar as the tearing of the chair was a result of its own tear-ability. It was naturally *gracious* insofar as this tear-*ability*, in unintentional relationship to Chris's labor (his nervous/pathological habit of scratching), provided him with a way out. But in what sense was this graciousness also the graciousness of "the spirits"?

The grace of the spirits is that grace which, as Corrington writes, invites us to an "unfolding of meaning," and to the "maximizing, under conditions of finitude," the efficacy of a sign, in this case, of cotton. It is noteworthy that the *Philosophy of Right*, the text in which Hegel describes the will of the slave as being sunk in its condition, was first published in 1821, just as the slogan "Cotton is King" was emerging to account for the growth of the American economy. The sovereignty of cotton was also then an economic-political sovereignty of the cotton industry over the will of the slave and a primary agent of black sunkenness.

Here, however, the cotton becomes the means by which Chris escapes sinking, insofar as it unexpectedly provoked him to think of a technological saving solution: namely, to think to stuff this cotton into his ears, thereby preventing the sound of the spoon scratching against the cup from reaching him, and thereby also providing him with a way out, just at the edge of the hypnotically induced Sunken Place. Chris found a way out. But this gift of the spirits, recognized as the gift of those spirits for whom cotton was apparently (and *only* apparently) the agent of their sunkenness, must also be thought in relationship to the etymological link which Martin Heidegger identified between thinking and thanking. Thinking and thanking have their etymological beginnings in the Old English *thanc*, with a "c."[21] To think is to thank, to have an original orientation of thankfulness toward that which provokes, or, to use his language, *calls*, one to think. In this case, it was the ancestral spirits, in gifting Chris cotton, who called Chris to think.

What I find crucial here is that this sacred fold of natural grace and the gift of the spirits, the cotton that called Chris to think, is irreducible to Chris's labor, to his pathological habit of scratching. The gift—the cotton—was given in excess of that labor insofar as, first, the tear-ability of the chair was/is a trait of the chair as a natural complex, regardless of his relationship to it, and, second, because on a semiotic and phenomenological level, the cotton *can only* be described as a gift of the spirits, of the ancestors over whom King Cotton was supposed to reign.

This giftedness of the cotton is, for me, extremely important and invaluable as an explosive semiotic potential, for us: it also calls *us* to think, thankfully, the irreducibility of the lives of those we call black to labor, even and especially in the everyday labor we call the *struggle*. It calls us to think the *grace* with which "black" lives, thankfully, escape reduction to the dialectical struggle for recognition between blackness and anti-blackness, reduction to a problem.

NOTES

1. W. E. B. Du Bois. *The Souls of Black Folk* (New York: Signet Classics, 2012), 7.
2. *Get Out*, Amazon Streaming, directed by Jordan Peele (Los Angeles: Monkeypaw Productions, 2017).
3. Ibid.
4. Ibid.
5. Ibid.
6. Ibid.
7. Ibid.
8. Georg Wilhelm Friedrich Hegel. *Philosophy of Mind: Part Three of the Encyclopedia of the Philosophical Sciences* (Oxford: Oxford University Press, 1971), 42, 43.
9. Ibid.
10. Hegel. *Elements of the Philosophy of Right* (Cambridge: Cambridge University Press, 1991), 37.
11. Ibid., 55.
12. Hegel. *The Encyclopedia Logic* (Indianapolis, IN: Hackett, 1991), 219.
13. Ibid., 220.
14. Ibid., 221.
15. Ibid., 87.
16. This is not the place to give a fuller articulation of this genealogy, but it would suffice, for now, to present a passage from a sixth-century chemical philosopher known as Pseudo-Olympiodorus. What is important about the following passage is that it reveals the articulation of a connection between blackness as sunken, as a product, as primitive, and, *qua* the internal relation with primitivity, of blackness as a phase of a *progressive* movement. This semiotic and conceptual space (blackness as a product, as primitive, and a phase of a progressive movement toward perfection) is what distinguishes the intentional structure of racial blackness from other kinds or conceptions of blackness. Here is Olympiodorus, as quoted by Oliver Dufault in "Transmutation Theory in the Greek Alchemical Corpus," *Ambix* 62, no. 3 (August 2015): 215–244: What we call bodies that are *drawn down* (*kataspao*) . . . are drawn down to (their) constitution and are turned into ashes (emphasis mine). And the fixating body/the body suitable for instantiation, which they call "black lead" . . . are the ashes and the residues of Maria. For they have known that these have existed from the beginning. Blackening happens because of this. During the practice, un-blackening, that is to say, whitening (happens because of this). For nothing else signifies whitening if not the discolouration by privation of the black. . . . For there are two colours that are diametrically opposed, white and black. White distinguishes but black comprehends. . . . Those who are

ignorant do not realize what distinguishes and what comprehends. What comprehends and what is much compacted is extracted from the appropriate body. . . . No longer revealing the invisible cosmos in itself, the soul reveals (itself) differently in another body of silver, and in silver, the fiery blood, that is to say, gold."

17. Elena Howe, "Seven Directors 'Weaponize Storytelling' to Stand Against Hatred and Violence," *Los Angeles Times*, November 30, 2017, https://www.latimes.com/entertainment/envelope/la-en-mn-directors-roundtable-20171129-story.html#nws=true.

18. Ibid.

19. Ibid.

20. Robert S. Corrington. *A Semiotic Theory of Theology and Philosophy* (Cambridge: Cambridge University Press, 2000), 212–224.

21. Martin Heidegger. *What Is Called Thinking?* (New York: Harper Perennial, 1976), 145.

REFERENCES

Bauer, Ralph. *The Alchemy of Conquest: Science, Religion*. Charlottesville: University of Virginia Press, 2019.

Corrington, Robert S. *A Semiotic Theory of Theology and Philosophy*. Cambridge: Cambridge University Press, 2000.

Du Bois, W. E. B. *The Souls of Black Folk*. New York: Signet Classics, 2012.

Get Out. Directed by Jordan Peele. Location: Los Angeles, 2017. Amazon, Streaming.

Hegel, Georg Wilhelm Friedrich. *Elements of the Philosophy of Right*. Cambridge: Cambridge University Press, 1991.

———. *Philosophy of Mind: Part Three of the Encyclopedia of the Philosophical Sciences*. Oxford: Oxford University Press, 1971.

———. *The Encyclopedia Logic*. Indianapolis, IN: Hackett, 1991.

Heidegger, Martin. *What Is Called Thinking?*. New York: Harper Perennial, 1976.

Howe, Elena. "Seven Directors 'Weaponize Storytelling' to Stand Against Hatred and Violence," *Los Angeles Times*, November 30, 2017, https://www.latimes.com/entertainment/envelope/la-en-mn-directors-roundtable-20171129-story.html#nws=true.

Newman, William. *Promethean Ambitions: Alchemy and the Quest to Perfect Nature*. Chicago: Chicago University Press, 2004.

Puett, Michael. "Ritual Disjunctions: Ghosts, Philosophy, and Anthropology," *The Ground Between: Anthropologists Engage Philosophy*, eds. Veena Das, Michael Jackson, Arthur Kleinman, and Bhrigupati Sing. Durham, NC: Duke University Press, 2014.

Sudan, Rajani. *The Alchemy of Empire: Abject Materials and Technologies of Colonialism*. New York: Fordham University Press, 2016.

Wennerlind, Carl. *Casualties of Credit: The English Financial Revolution*. Cambridge, MA: Harvard University Press, 2011.

Chapter 5

Racism, Religious Education, and Transformation

Moon Son and Ji Young Park

INTRODUCTION

This chapter reflects the regional context of Northeast Asian countries embodied in the transformative contexts such as the Covid-19 pandemic and the nuclear tension between North Korea and the United States.[1] On the basis of the first discovery of Covid-19 from Wuhan, China, U.S. president Donald Trump referred to the coronavirus disease as "the Chinese virus" and such misunderstanding was linked to the reinforcement of "racial discrimination" with certain types of "foreign policy and trade negotiations."[2] This shows well that health threats such as the Covid-19 pandemic may affect the misappropriation of racial and ethnic discrimination leading to "division and fear of others."[3] Here we use a methodological inquiry based in practical theology to analyze contemporary political affairs and offer creative responses within the field of religious education. The ecology of religious education can help to dismantle the threat of ethnic and racial discrimination such as white supremacy and cultural imperialism. The method of religious education developed here supports a shared pedagogy emerging between students and their teachers in the narrative context of Jesus's challenge to all forms of oppression as the radically democratic presence of God.

This is similar to the sign system of yin-yang in Eastern thought. Even though *yin* and *yang* have the opposite characteristics and resisting colors, they transcend each other and transform into the other.[4] The harmonious relation of yin yang therefore symbolizes the coexistence of resisting oppositeness. This means the importance of religious education to dismantle racial discrimination and cultural prejudice. Especially we consider an ecstatic transformation between *nature naturing* and *nature natured* to be expressed by the communicative consciousness of selving process in nature and its innumerable orders of the world suggested by Robert S. Corrington, an American philosopher who developed "ecstatic naturalism."[5] This means to transcend the boundaries between the self and others and approach the encompassing of the self-othering through the interactive relationship between *nature naturing* and *nature natured* of our selving process. Such selving process of our communicative consciousness may encourage the dismantling of racial and ethnic discrimination in an ecology of religious education.

As the forty-fifth president of the United States (2017–2021), Donald Trump's "fire and fury" remark toward the North Korean regime provoked an intensification of the military tension on the Korean Peninsula and triggered the North Korean nuclear threat to the U.S. territories through the frequent testing of nuclear weapons and intercontinental ballistic missiles that could arrive in Hawaii or arrive at the West Coast of the United States. A central ethical aim of this research is a basic ideal to see the diplomatic circuit currently existing between the North Korean dictator Kim Jong-un and Donald Trump to swing from the North Korean nuclear threat and its counterstrike from the United States toward a more comprehensive perspective of practical theology rather than the current Western imperialist rules based in white supremacy that are the current political system in Washington, DC. And this research pays special recognition to the potential comprehensiveness of practical theology to be applied in the political discourse about the ongoing nuclear issues between the United States and North Korea. Such comprehensive flexibility of practical theology will be served as the academic and practical framework to exchange the political discourse of the "Washington consensus" for a more often and effective dialogue of intercultural religious education.

Bruce W. Jentleson, who is a professor of public policy and political science at Duke University, notices that Donald Trump's foreign policy as the "America First" approach globally makes regional issues in Asia much more problematic.[6] He argues that Trump's economic and assertive nationalism will make the Chinese premier Xi Jinping take the position of "the champion of free trade" among Asian countries.[7] And this entails a quantum loss of the U.S. global influence in these regions. Furthermore, through the military alliances between North Korea, China, and Russia in Northeast Asia, there is a high-risk factor that the United States may lose the golden time to protect very serious threats of North Korea's nuclear program and intercontinental missile systems.[8]

As Asian religious scholars encountering these imminently foreboding nuclear threats as a daily reality, we experience directly and existentially the seriousness of Trump's reckless foreign policy as the arrogant imperialism of an "America First" approach is cut from the same cloth of ideologies of "white supremacy." That is a survival issue and reality of many ethnic and racial generations in such Northeast Asian contexts and the Trump foreign policy is nothing short of genocidal. The sixty-sixth U.S. secretary of State, Dr. Condoleezza Rice, has remarked, "North Korea's nuclear program was a global, not just a regional, issue. Its treatment of its own people offended not just the President but also our country's commitment to human rights" and we here will also proceed from a basic attitude and theoretical stance to approach the issue of North Korea's nuclear program as part of a foundational principle of democracy involving human rights, rather than an assertive nationalism as a pseudo-alternative to the ideology of white supremacy.[9]

As religious educators in a Northeast Asian context, we draw upon a productive dialogical common ground to be found between political affairs and religious education from George Albert Coe's memorable sentence: "Religious education alone takes account to the whole personality, of all its powers, all its duties, all its possibilities and of the ultimate reality of the environment."[10] That has the distinct feature of the twenty-first century's reframing structure of progressive education to include the practice of social engagement for reconstructing society and government in the meaningful purpose of religious education.[11] The main concern of my research here is the dialectical ecology of religious education and its potential to undo the racial and ethnic processes of white supremacy and to encourage instead a dialogic ethical "principle of reciprocity" among participants as the ideals of democracy have to realize "the socioethical ideal of the democracy of God" on the basis of the Christian faith.[12]

Drawing especially upon the frame of reference of "White Practical Theology" suggested by Tom Beaudoin and Katherine Turpin, we might use this as a theoretical scaffolding setting of practical theology for a renewed "methodological inquiry" to analyze the "cultural privilege of whites" and the "white discipline" embodied in Trump's foreign policy as the America First approach.[13] Beaudoin and Turpin's methodological frame is a powerful reference of practical theology to put into stark contrast the academic and practical inconsistencies between political affairs and religious education.

In addition, we understand their methodology of practical theology as an adequate formulation to be dismantling the "cultural homogeneity" of white supremacy and instead universalizing an ecology of religious education that can undo the threat of "race and racism" and to form a more expansive and inclusive "religious identity" for Asian people in their own cultural contexts.[14] In other words, the ecology of religious education to encourage personal and corporate theological reflection in order to dismantle the threats of racial and ethnic discrimination of white supremacy recognizes brave challenges of our learners and their teacher's endeavors to open the expansion of their religious identity formation in a shared pedagogy between students and their teachers in light of "the narrative of Jesus to challenge all forms of oppression."[15]

THE OTHER SIDE OF THE WORLD: CHINA AS AN UPRISING EAST ASIAN COUNTRY

The regional issues in religious education of course have unique meanings. A representative book that deals with the regional issues of religious education is Richard Osmer and Friedrich Schweitzer's *Religious Education between Modernization and Globalization: New Perspectives on the United States and Germany*. In that book, Osmer and Schweitzer describe the transformative progress of these two countries and how the United States and Germany operate in a transitional era from modernization to globalization. In other words, the regional issue of both countries constitutes a main focus of their theoretical outlook on geopolitical situations. They emphasize that "religious education stands in interdependent relationships to its social contexts" and that the main goal of religious education is closely related to "a common goal of promoting the well-being of our societies and the human community as a whole."[16]

In this aspect of a truly global and multicultural theory of religious education, we too have recently become quite interested in the regional issues occurring between Korea and China. The relationship between both countries adds significantly to the dynamics of such meanings and conspires with the relational issues simultaneously occurring between the United States and Japan in the context of the Korean Peninsula. For example, Chinese students now make up the largest percentage of foreign students studying in Korean universities. And very recently it is estimated that approximately 800,000 Korean people are living in China. According to a 2007 statistical analysis of Korean World Mission Association, there were 14,905 Korean Protestant missionaries working in 168 countries and among them, Korean missionaries working in China were recorded at the first rank with 2,640 missionaries among them.

The dual issues of joining the Asian Infrastructure Investment Bank (AIIB) led by China and the United States' offensively deploying the so-called Terminal High-Altitude Area Defense (THAAD) system in the Korean Peninsula have promoted the urgent need to have more multilateral and communicative relationships between Korea, China, and the United States. In fact, many Koreans have expressed many concerns about Chinese foreign leadership under the auspices of Chinese president Xi Jinping. China and the United States are the two superpowers in foreign and economic affairs. And in the midst of all the geopolitical tensions between China and the United States and between China and Japan, President Xi has been strenuously carrying out an "anti-corruption campaign" as a paramount Chinese domestic agenda.[17] Moreover, his personal confidence is high in China and among the citizenry of other nations. Thus, many people anticipate an increasing need to act cautiously toward the questions of putting pressure on China in the name of Chinese democratization.

However, professional political scientists also recognize that President Xi remains firmly within the stereotypical pattern of "a pragmatic Chinese leader" and has not really been substantially changing "the course of China's grand strategy of a 'peaceful rise'" which is shaped by foreign policy of Chinese "self-righteousness" to maintain a real "competitor" in relation to the United States and world affairs.[18] For instance, Chinese political scientist Zhu Feng, who is professor of International Relations at Nanjing University, sees the "territorial disputes in the East and South China Seas" within the context of "China's military modernization" and its expansion in relation to the concrete realization of a "Chinese 'self-righteousness' mindset" rather than as representing a direct challenge to the U.S. central role as the police of a neoliberal world order, and thereby signaling the emergence of a "new Cold War."[19]

Since Xi Jinping's political confidence is very high, it is not easy to anticipate any transformative changes in Chinese political affairs, as the Chinese political leadership maintains its traditional pattern of a largely self-enclosed framework. The U.S. influential political scientist, Susan Shirk, who is a professor of China and Pacific relations at UC San Diego and former deputy assistant secretary of State under the Clinton administration, analyzes the oligarchical structure of the Chinese Communist Party (CCP). Because the leadership election process is very conservative and closed to public proceedings, generally speaking, in China, the political decision-making is determined by the Politburo Standing Committee and/or the Military Commission. According to Susan Shirk's political analysis, Chinese "leadership competition" is shrouded by various "wraps."[20] For example, the hottest political issue of "the Seventeenth Chinese Communist Party Congress in fall 2007" was who would become a successor of President Hu Jintao.[21] While many people thought it would be Zeng Qinghong, who was the right-hand man of former president Jiang Zemin, the Central Military Commission did not appoint Zeng, who served as vice president, in its number two position, since he was very old.[22] Of course, we are able to recognize that China currently experiences an "explosion of information" through a proliferation of "market-oriented and Internet-based new sources."[23] However, the CCP still largely controls the public's access to media and information through its strong digital regulation devices and monitors "the actions of subordinate officials" in the local governments through recognizing the openness of media "as a watchdog."[24]

In such current China-focused circumstances, it is important how we go about constructing effective projects of transformative religious education to apply to the foreign relations predicaments occurring between the United

States, China, and South and North Korea. Here we are exploring the potentiality that cultural heterogeneity has within the context of developing a theory and practice of religious education in the unique perspective of each of these countries' resources used to be developing foreign policy.

THE COMMUNICATION WITH HETEROGENEITY: THE NARRATIVE OF *THE GREATEST SHOWMAN*

In the film *The Greatest Showman*, Barnum experienced big despair in that his circus performance building was set fire by protestors. We want to show how this film from the perspective of a theory of communication can be used to highlight the ethical role of heterogeneity in intercultural communication. This tragic part of the movie shows the success of communication between circus members and their leader. The success of communication means to arrive at some consensus in their thoughts and ideas.[25] We think that a degree of consensus among the members of that community makes a decision whether or not their communication succeeds. Even though some protestors broke the place for their show, in the scenes which we have seen, they do not fail in their communication. Such behavior means the successful exchange of their ideas or thoughts.

In other words, the status of interdependence among the members of that community was conceived as very positive. Barnum's community maintained a series of "interacting components."[26] And here we can recognize that a series of dynamic communicative mechanisms and normative structures were regulating each of the circus members' behavior, even though their community's heterogeneity was very high. This means that there was a larger system to include and encompass Barnum's communicating community. What was this larger system to encompass the provincial Barnum's circus community? We believe it was the sheer diversity as relational strength and creative, communicative democracy as religious education. Our generation recognizes and appreciates the radical heterogeneity which the narrative of *The Greatest Showman* emphasizes. In our minds, Barnum's system exists in our generation's perspective to embrace such difference and diversity. And it is precisely because of heterogeneity that such a communicative system's sustainability can be maintained and enhanced in ongoing interpretations.

In this movie, we can find a way to manage and cope with the conflict between Barnum's members. Any effective communication pattern can resolve some differences effectively. However, sometimes the continuity of entrenched habits of communication may deepen the conflicts between members, since their differences are overemphasized. In the context of this movie we see that the heterogeneity successfully gives rise to deepening solidarity of their emergent community.[27] Here such heterogeneity can be understood in the perspective of a new potentiality for a community facing similar problems (i.e., a shared global predicament and the emergence of a new planetary public). We can recognize that any heterogeneity can motivate the potentiality of a community to enhance its communicative productivity as ethical-religious educational outcomes that expand and enrich the human self and its manifold of community belongings through a context of global religious education.

THE CONDITION OF THE SELF FOR HUMAN DEVELOPMENT

The formation of a human self to be effective in sacred and secular worlds is a relatively recent concern of religious education. Human developmental thought is always a natural phenomenon. In previous work in the psychological studies of the process of human development the concept of post-conventionality has been used to indicate a highest stage of ethical human development. Modern people typically go through a process of development from the pre-conventional stage through the conventional period to the postconventional level. When we study the process of human development, we should focus more on the ecstatically transforming process of selving. Here, human development always means the growth of the ecstatic self as a potential contributor to any number of communicating communities. In other words, that means the development in the structure of the self, that is, the structure of human communicative consciousness.

How does such human development and the Christian Trinity connect with each other? We can find the possible relationship between human development and the Trinity in the Bible: "So God created humans to be like himself; he made men and women."[28] This biblical scripture shows the structural similarity between human development and the Trinity. The social doctrine of the Trinity is expressed well in how mutual indwelling and reciprocal interpenetration based on the principles of cooperation and communicative equality occurs between three Persons: God the Father, God the Son (Jesus Christ), and the Holy Spirit. The feminist theologian Sallie McFague argued that the traditional names of the Trinity have been very male-oriented. Hence, she suggests that we use more gender neutral or inclusive words such as the following: Creator, Redeemer, and Sustainer instead of the Father, the Son, and the Holy Spirit.[29] Likewise, the social doctrine and the feminist perspective of the Trinity reflect well the principles of cooperation and equality in the divine entity as communicative advance. At the same time, these principles of the Trinity provide the meaningful insights for human development. This is the transforming potential as communicative process in a triadic structural pattern of human consciousness.

Robert Kegan focuses upon this structural pattern of human consciousness. He argues that human consciousness experiences five stages in its developmental pattern and has a helix shape in the sensitive, interpenetrating, mutual relationship between subject and object. The shape of a helix means the dynamic transition or transformation between the human developmental process and her communicative lifeworld. The first order of human consciousness is that of children between the ages of two and six. Their thinking is fantastic and illogical, their feelings are impulsive and fluid, and their social relations are egocentric.[30] Next comes the stage of children between the ages of seven and ten which tends to have the characteristics of the second order of human consciousness. Here thinking becomes more concrete and logical, and their feelings have time-enduring characters, and their social relations reflect "a separate mind and a distinct point of view."[31] The third order of human consciousness reflects the psychological characteristics of adolescents—their thinking is more abstract, and their feelings tend to show more self-reflexive emotion, and their social relations reveal the loyalties and specific devotions they might have developed toward people in the community or some ideas.[32]

The adults in the fourth order of human consciousness have "a self-evaluating, self-governing system"[33] and their consciousness expresses "a complex or integrated system."[34] These fourth-order adults tend to emphasize "the continuous creating and recreating of roles" rather than the strict adherence to our given social roles.[35] The fifth order of consciousness has a quality of the multi-self which is always pluralistically nuanced, with multiple and interrelated configurations of worldviews. In this highest order of consciousness, some people recognize the coexistence of difference and conflicts between distinct peoples, groups, communities, and societies. Kegan applies such coexistence of difference and conflicts within human mind to a kind of "conflict resolution" strategy available in transformative society. He describes the implementation of this strategic, higher-order conflict resolution as imaginative role-taking or reversing settled, affective dispositions in the following way: "Postmodernism suggests a kind of 'conflict resolution' in which the Palestinian discovers her own Israeli-ness, the rich man discovers his poverty, the woman discovers the man insider her."[36]

The transition from the first order to the second order can be realized by the durability of a singular human self. Children between the ages of seven and ten can typically sustain more durable characters in their thinking, feeiing, and social relation than children between the ages of two and six. The transition from the second order to the third order also means the expansion of ethical scope where the self of the third order can use more comprehensive resources such as thinking, feeling, and social relation to constitute a communicating human consciousness higher than that of the second order. For example, adolescents usually have broader comprehensibility in thinking, feeling, and social relations than elementary school students. The fourth order and the fifth order of human consciousness similarly divide the development of adult person into two contrasting types of characteristics. The subjective character is remarkable in the fourth order and the intersubjective or multi-subjective communicating character is more influential in the fifth order. The fourth order of human consciousness has a persistent ego boundary. While the fifth order of human consciousness accepts the place for others within a more fluid ego boundary, it entails the communicative coexistence and mutual transformation of self and others in an encompassing worldview. Therefore, in the fifth order of consciousness, the thinking, feeling, and social relations with paradoxical patterns embracing heterogeneity as an ethical resource and communicative keynote become possible.

The Trinity also has such a paradoxical or "absurd" pattern. The coexistence of the one and many is always possible in any of the Three Persons of the Trinity. Three Persons in the Trinity have their distinguishing character traits as the Father or Creator, the Son or Redeemer, and the Holy Spirit or Sustainer. However, only these three Persons in communicative agape can sustain the dynamic, triadic unity. Here, we can find a semiotic resource to think through the coexistence of unity and within and through diversity. Sometimes, when we encounter new life circumstances, we experience severe anxiety or traumatic embarrassment. But, our selves always have the mysterious power over our limited consciousness. This means that our selves encounter the transcendental God in the "orders of nature" of "the selving process." In this aspect, Jea Sophia Oh, an Asian ecofeminist theologian and a comparative philosopher, states that we can find "the selving process through the order of nature" which is "nurtured from a source outside of itself."[37] Furthermore, Corrington writes, "Selving is a gift of that dimension of God that drives toward the future and the manifestation of greater actuality and consciousness."[38] Hence, through the selving process, we can discover a surprising capacity to overcome the negative barriers of our circumstances as a gift of transcendence. As Corrington calls selving "the ongoing process of self-formation,"[39] in terms of Christian education as a pedagogy for selving process, we believe that this research can help teachers and students realize and vitalize this very mysterious capability in classrooms, which can in turn shape the larger cultural communicative patterns that make possible the very conditions for the emergence of foreign policy in an increasingly perilous age of globalization.

DESIGNING THE TRANSFORMATION OF HUMAN SELF IN THE ECOLOGY OF RELIGIOUS EDUCATION

Human selving demands for some radical transformations in its substantive and processual encounter with the world. The meaning of curriculum can be understood as the cultural contents which human self in the world is encountered with. Kegan distinguishes the modern self and the transformative self through distinctive ways to construct and cope with "difference and conflict."[40] A typical modern self in the comparative relation to a transformative self has the completeness and wholeness of the meaning of its worldview—that is one's "*own* way of looking at and relating to the world" (emphasis in original). [41] Any person who has such a working conception of modern selfhood typically does not try to change another person's perspective. Instead they engage in collaborative communicative practices that together seek to explore a more comprehensive view in order to solve problems of difference and conflict. This is the ethical attitude and communicative strategy of the modern self in a nutshell.

The transformative self has a very different stance regarding the resolution of differences and oppositions within modernity. Usually, we are used to finding differences and oppositions in the other person rather than in our own selves. However, the transformative self seeks to understand differences and oppositions in the other person as a part of our whole narrative process of always constructing anew our concrescent ethical plurisinguarity. In other words, the transformative self utilizes differences and oppositions with the other person as some chance to be always recovering "our own complexity."[42] Likewise, the transformative self has a profound capacity to lead us to see "our own oppositeness" within difference and conflict with other persons.[43]

The transformative self can signal a new communicative-organism of "trans-system" or cross-forming to move a core of a static self from reified subject-object to ecstatic communicative agent.[44] Such dynamics can be understood as an exercise in ethical "multipleness" rather than systemic "completeness."[45] The transformative self then has some dynamic power to "relativize *systemic knowing*" allowing us to move beyond the current limitations of our wholeness (emphasis in original).[46] A transformative curriculum therefore must distinguish between normative goals and descriptive goals. A normative goal depends on the wholeness of the curriculum of the modern world. But a descriptive goal recognizes the multiple possibilities of a curriculum to be transforming worldviews. Therefore, adult learners need to have "self-direction" and need to be challenging "the assumptions of wholeness, completeness, and the priority of the [fixed] self" in the encounter with any transformative curriculum in religious education. In the end, the relation with other persons transforms a single system of our self into a communicative manifold of "our multiple selves" that calls for a deep recognition of our own limitations.[47]

This transformative model for expanding human selving as communicative process in religious education may be further enhanced and understood drawing upon Mary Elizabeth Mullino Moore's model of relational teaching. Her relational teaching as narrative method is precisely explained in her *Teaching from the Heart: Theology and Educational Method*. In this book, she describes the narrative method as the deployment of certain "passions" to connect person and event, cultural and religious stories, and the stories of person and the earth.[48] Moore recognizes the educational meaning of narrative teaching in seeing human life in the holistic relationship between "the parts" and "the whole."[49] To see the parts in relation to the whole means to understand our lives in theological terms and in the process of the faithful reorganization of experience through the biblical narratives. She also emphasizes the importance of narrative teaching as "a significant mode of human communication."[50] Moreover, Moore supposes the organic and weblike interconnected view of the world. It becomes the precondition of narrative communication regarding the relationship between the world and human beings. By the same token, narrative teaching has considerable capacity to generate educational and "moral imagination" in our daily lives and in many Christian communities, rather than defaulting to the flattening "cause-and-effect way" of interpreting conflict.[51]

The role of imagination in narrative teaching is to provide learners with "artistic experience."[52] And educational imagination promotes a deep and primitive perspective to connect the human "ecstatically" with her natural words.[53] This means the profound linking of theology and imagination through stories. Moore understands the main distinction of narrative thought in some natural connection between action and consciousness. Here consciousness means comprehensively all of human thinking, feeling, and willing. She emphasizes that narrative thought is different from "logico-scientific thinking" grounded as it is in a reductive epistemic register of "description, explanation, and verification."[54] Therefore, narrative teaching promotes a sustainable and organic link between personal consciousness and social reality as communicating communities, since human consciousness reflects imaginative and individual awareness and experience of the world with "a form of indirect communication."[55] And Moore understands the importance of narrative teaching as an essential channel for human growth and the initiative dynamic to form and transform the world.[56] Likewise, narrative method generates the transformation of human consciousness and the reality of the world through both organic relations of human and their worlds with a mode of indirect communication that promotes artistic experience for learners. It is especially vitally important that we need to know the role of narrative teaching in addressing a sensitive tension or conflict that might be generated between religious and cultural differences in public spaces. In this aspect, Moore's educational model can be applied as the device of religious education to transform the distinctive resources of heterogeneity and difference into a new potentiality to be reformulating human selving in our public and secular societies where the harshness of racism, xenophobia, militarism, and imperialism are all-too-humanly-present.

CONCLUSION

We recently read a report that "President Donald Trump said Tuesday he's not looking to revive the much-criticized practice of separating migrant children from their families at the southern border. At the same time, he suggested the policy had worked to deter migrants from coming into the United States, although he offered no evidence to support his position"—this article shows that the *topos* of space is closely related to the issue of racism.[57] The Covid-19 pandemic also focuses on the space deterrence of health policies such as self-isolation and social distancing, and such systems would lead to racialized and discriminatory debates to affect marginalized groups.[58] We recognize that Corrington's hermeneutical world of "naturing nothingness" to eliminate "false plentitude and the carver and shaper of forms of betweenness" has pragmatic utility to construct an ecology to diminish the issue of space in racism.[59] It brings to mind James Fowler's idea of "conjunctive faith" to be the clearing away of the line "between 'us' and 'them'" to reconstruct our common symbol and myth in light of radical decentralizing love of Jesus Christ.[60] Furthermore, the transcendental "traits" of naturing nothingness to move beyond the spatial boundary of "Being" as "*within* and *without*" may provide the potentiality to liberate the self in a closed universe.[61] Especially, the encompassing mode of nothingness as a "presence of God" may guide the "road to freedom" to dismantle the ego boundary working as the initial point of all spatial traits and a hard foundation of racism in light of the "Spirit of God."[62] Therefore, we naturally reconsider Jesus's statement as the principle to overcome racism:

> I tell you for certain that a grain of wheat that falls on the ground will never be more than one grain unless it dies. But if it dies, it will produce lots of wheat. If you love your life, you will lose it. If you give it up in this world, you will be given eternal life.[63]

When we regard the tasks of Christian educators as the activities to teach and interpret the perspectives of Christian faith to the wider world, Christian believers' responsibility for dismantling racism can be understood in the transcendental movements to help the "least" of our societies remained in "limited life chances" to be liberated from their enslaved border barriers in the public sphere.[64]

NOTES

1. Revised version of Moon Son and Ji Young Park's "Creating Change in the Ecology of Religious Education," *Journal of Christian Education in Korea* 61 (2020): 109–129. Printed with permission.

2. Delan Devakumar, Geordan Shannon, Sunil S. Bhopal, and Ibrahim Abubakar, "Racism and Discrimination in COVID-19 Responses," *Lancet* 395, no. 10231 (2020): 1194. doi: 10.1016/S0140-6736(20)30792-3.

3. Ibid.

4. Jea Sophia Oh, "Vulnerable Transcendence of Nature: A Naturalistic Reading of Hybridity, Beginning, and Colliding in Chinese Creation Mythology," In *Nature's Transcendence and Immanence: A Comparative Interdisciplinary Ecstatic Naturalism*, ed. Marilynn Lawrence and Jea Sophia Oh (New York: Lexington Books, 2018), 130.

5. Robert S. Corrington, *Deep Pantheism: Toward a New Transcendentalism* (Lanham, Maryland: Lexington Books, 2016), x.

6. Bruce. W. Jentleson, "Trump's Global Foreign Policy Is Bad News for Asia," *Global Asia* 12, no. 4 (2017): 10.

7. Ibid., 11.

8. Miles Kahler, "Asia's Trump Peril: Reckoning with Economic Conflict," *Global Asia* 12, no. 4 (2017): 27.

9. Condoleezza Rice, *No Higher Honor: A Memoir of My Years of Washington* (New York: Crown, 2011), 37.

10. George Albert Coe, "Religious Education as a Part of General Education," In *The Religious Education Association: Proceedings of the First Convention, Chicago 1903*, 44–52 (Chicago: The Religious Education Association, 1903), http://old.religiouseducation.net/journal/historical/Coe.pdf.

11. Richard Robert Osmer and Friedrich Schweitzer, *Religious Education between Modernization and Globalization: New Perspective on the United States and Germany* (Grand Rapids, Michigan: William B. Eerdmans, 2003), 84.

12. Diane L. Moore, *Overcoming Religious Illiteracy: A Cultural Studies Approach to the Study of Religion in Secondary Education* (New York: Palgrave Macmillan, 2007), 11; Osmer and Schweitzer, *Religious Education between Modernization and Globalization*, 87.

13. Tom Beaudoin and Katherine Turpin, "White Practical Theology," in *Opening the Field of Practical Theology: An Introduction*, ed. Kathleen A. Cahalan and Gordon S. Mikoski (New York: Rowman & Littlefield, 2014), 259.

14. Beaudoin and Turpin, "White Practical Theology," 267; Joyce Ann Mercer, "A Space for Hard Conversation on Race, Racism, Anti-racism, and Religious Education (Editorial)," *Religious Education* 112, no. 1 (2017): 2.

15. Willie James Jennings, "Race and the Educated Imagination: Outlining a Pedagogy of Belonging." *Religious Education* 112, no. 1 (2017): 64; Joyce Ann Mercer, "Feminist and Womanist Practical Theology," in *Opening the Field of Practical Theology: An Introduction*, ed. Kathleen A. Cahalan and Gordon S. Mikoski (New York: Rowman & Littlefield, 2014), 98.

16. Osmer and Schweitzer, *Religious Education between Modernization and Globalization*, 5.

17. Zhu Feng, "Geopolitics and China's Response: Be a Co-operator and Competitor," *Global Asia* 9, no. 3 (2014): 27.

18. Ibid.

19. Ibid.

20. Susan L. Shirk, *China: Fragile Superpower* (New York: Oxford University Press, 2008), 52.
21. Ibid.
22. Ibid.
23. Susan L. Shirk, "Changing Media, Changing China," in *Changing Media, Changing China*, ed. Susan L. Shirk (New York: Oxford University Press, 2008), 2.
24. Ibid., 5.
25. Stephen W. Littlejohn, Karen A. Foss, and John G. Oetzel, *Theories of Human Communication*, 11th ed. (Long Grove, IL: Waveland, 2017), 4.
26. Ibid., 42.
27. Ibid., 240.
28. Gen. 1:27 (Contemporary English Version).
29. Catherine Mowry LaCugna, "God in Communion with Us," in *Freeing Theology: The Essentials of Theology in Feminist Perspective*, ed. Catherin Mowry LaCugna (New York: Haper Collins, 1993), 105.
30. Robert Kegan, *In Over Our Heads: The Mental Demands of Modern Life* (Cambridge, MA: Harvard University Press, 1994), 29.
31. Ibid.
32. Ibid., 32.
33. Hyun-Sook Kim, "Christian Education for Postconventional Christian Identity in the Modern Context," *International Journal of Practical Theology* 5, no. 2 (2001): 275.
34. Kegan, *In Over Our Heads*, 92.
35. Ibid., 96.
36. Ibid., 320–321.
37. Oh, "Vulnerable Transcendence of Nature," 132.
38. Robert S. Corrington, "Divine Nature," in *Nature's Perspectives: Prospects for Ordinal Metaphysics*, ed. Armen Marsoobian, Kathleen Wallace and Robert S. Corrington (Albany, NY: SUNY Press, 1991), 354.
39. Robert S. Corrington, *Nature's Sublime: An Essay in Aesthetic Naturalism* (Lanham, MD: Lexington Books, 2013), 29.
40. Kegan, *In Over Our Heads*, 307.
41. Ibid., 308.
42. Ibid., 310.
43. Ibid.
44. Ibid., 313.
45. Ibid.
46. Ibid., 317.
47. Ibid., 351.
48. Mary Elizabeth Mullino Moore, *Teaching from the Heart: Theology and Educational Method* (Harrisburg, PA: Trinity Press International, 1998), 131.
49. Ibid.
50. Ibid., 132.
51. Ibid., 138.
52. Ibid., 139.

53. Ibid.
54. Ibid., 140.
55. Ibid., 141.
56. Ibid., 142–143.
57. *The Korea Herald*, April 11, 2019.
58. Devakumar et al., "Racism and Discrimination in COVID-19 Responses," 1194.
59. Robert S. Corrington, *Nature and Nothingness: An Essay in Ordinal Phenomenology* (Lanham, Maryland: Lexington Books, 2017), 125.
60. James W. Fowler, *Becoming Adult, Becoming Christian: Adult Development and Christian Faith* (San Francisco, CA: Jossey-Bass, 2000), 54.
61. Corrington, *Nature and Nothingness*, 125–126.
62. Jean Vanier, *Becoming Human* (New York: Paulist Press, 1998), 122–123.
63. John 12:24–25 (Contemporary English Version).
64. Jack L. Seymour, *Teaching The Way of Jesus: Educating Christians for Faithful Living* (Nashville, TN: Abington Press, 2014), 31–33.

REFERENCES

Beaudoin, Tom, and Katherine Turpin. "White Practical Theology." In *Opening the Field of Practical Theology: An Introduction*, edited by Kathleen. A. Cahalan and Golden S. Mikoski, 251–69. New York: Rowman & Littlefield, 2014.

Coe, George Albert. "Religious Education as a Part of General Education." *The Religious Education Association: Proceedings of the First Convention, Chicago 1903*, 44–52. Chicago: The Religious Education Association, 1903. http://old.religiouseducation.net/journal/historical/Coe.pdf.

Corrington, Robert. S. *Deep Pantheism: Toward a New Transcendentalism*. Lanham, MD: Lexington Books, 2016.

———. "Divine Nature." *Nature's Perspectives: Prospects for Ordinal Metaphysics*, edited by Armen Marsoobian, Kathleen Wallace and Robert S. Corrington. Albany, NY: SUNY Press, 1991.

———. *Nature's Sublime: An Essay in Aesthetic Naturalism*. Lanham, MD: Lexington Books, 2013.

———. *Nature and Nothingness: An Essay in Ordinal Phenomenology*. Lanham, MD: Lexington Books, 2017.

Devakumar, Delan, Geordan Shannon, Sunil S. Bhopal, and Ibrahim Abubakar. "Racism and Discrimination in COVID-19 Responses." *Lancet* 395, no. 10231 (2020): 1194–1194. doi: 10.1016/S0140-6736(20)30792-3.

Feng, Zhu. "Geopolitics and China's Response: Be a Co-operator and Competitor." *Global Asia* 9, no. 3 (2014): 24–29.

Fowler, James W. *Becoming Adult, Becoming Christian: Adult Development and Christian Faith*. San Francisco, CA: Jossey-Bass, 2000.

Jennings, Willie James. "Race and the Educated Imagination: Outlining a Pedagogy of Belonging." *Religious Education* 112, no. 1 (2017): 58–65.

Jentleson, Bruce W. "Trump's Global Foreign Policy is Bad News for Asia." *Global Asia* 12, no. 4 (2017): 10–14.
Kahler, Miles. "Asia's Trump Peril: Reckoning with Economic Conflict." *Global Asia* 12, no. 4 (2017): 26–31.
Kegan, Robert. *In Over Our Heads: The Mental Demands of Modern Life*. Cambridge, MA: Harvard University Press, 1994.
Kim, Hyun-Sook. "Christian Education for Postconventional Christian Identity in the Modern Context." *International Journal of Practical Theology* 5, no. 2 (2001): 260–286.
LaCugna, Catherine Mowry. "God in Communion with Us." *Freeing Theology: The Essentials of Theology in Feminist Perspective*, edited by Catherin Mowry LaCugna, 139–60. New York: HaperCollins, 1993.
Littlejohn, Stephen. W., Karen. A. Foss, and John G. Oetzel. *Theories of Human Community*. 11th ed. Long Grove, IL: Waveland Press, 2017.
Mercer, Joyce Ann. "A Space for Hard Conversation on Race, Racism, Anti-racism, and Religious Education (Editorial)." *Religious Education* 112, no. 1 (2017): 1–2.
———. "Feminist and Womanist Practical Theology." In *Opening the Field of Practical Theology: An Introduction*, edited by Kathleen A. Cahalan and Gordon S. Mikoski, 97–114. New York: Rowman & Littlefield, 2014.
Moore, Diane L. *Overcoming Religious Illiteracy: A Cultural Studies Approach to the Study of Religion in Secondary Education*. New York: Palgrave Macmillan, 2007.
Moore, Mary Elizabeth Mullino. *Teaching from the Heart: Theology and Educational Method*. Harrisburg, PA: Trinity Press International, 1998.
Oh, Jea Sophia. "Vulnerable Transcendence of Nature: A Naturalistic Reading of Hybridity, Beginning, and Colliding in Chinese Creation Mythology." In *Nature's Transcendence and Immanence: A Comparative Interdisciplinary Ecstatic Naturalism*, edited by Marilynn Lawrence and Jea Sophia Oh, 129–36. New York: Lexington Books, 2018.
Osmer, Richard Robert, and Friedrich Schweitzer. *Religious Education between Modernization and Globalization: New Perspective on the United States and Germany*. Grand Rapids, Michigan: William B. Eerdmans Publishing Company, 2003.
Rice, Condoleezza. *No Higher Honor: A Memoir of My Years of Washington*. New York: Crown Publishers, 2011.
Seymour, Jack L. *Teaching The Way of Jesus: Educating Christians for Faithful Living*. Nashville, TN: Abingdon Press, 2014.
Shirk, Susan L. "Changing Media, Changing China." In *Changing Media, Changing China*, edited by Susan L. Shirk, 1–37. New York: Oxford University Press, 2011.
———. *China: Fragile Superpower*. New York: Oxford University Press, 2008.
Vanier, Jean. *Becoming Human*. New York: Paulist Press, 1998.

Chapter 6

A Phenomenological Study of Feminist Political Consciousness

Susan Erck

INTRODUCTION

What are the psychological barriers to attaining feminist political consciousness? How are these barriers overcome? These questions are of interest in feminist philosophy as well as system justification theory. This chapter seeks to answer these questions in novel ways by taking the system justification of religious violence as analogous to system justification of violence against women and girls. I employ Robert Corrington's existential, ordinal phenomenology of psychosocial processes involved in religious violence. I invoke Christine de Pizan's *The Book of the City of Ladies* for a classic feminist depiction of the set of experiences to be phenomenologically described.

SYSTEM VERSUS SELF

System justification theory attempts to answer the puzzling question, "Why do the disadvantaged sometimes participate in their own subordination?"[1] Why, for example, do women sometimes hold sexist attitudes about the characteristics and capabilities of their own sex? Social psychology has formulated explanations for the presence of sexist attitudes in men through *social identity theory*, which posits an ego justification and group justification psychological motive for males to view themselves as the superior sex.

Yet this motive cannot explain the presence of sexist attitudes in females. John Jost attempts to fill this explanatory gap with the idea that individuals not only have an interest in holding a favorable attitude about themselves (ego justification) and groups to which they belong (group justification) but are also motivated to view the *social system* in which they live as fair and just.

Those who are disadvantaged will nonetheless hold certain ideologies to be true which justify the system disadvantaging them, at the expense of their ego and group justification—for example, vindicating the system at the expense of their lowered self-esteem and a depressed sense of entitlement. System justification theory has drawn inspiration and insight from Marxist-feminist theory of "false consciousness," examining the psychology behind women's lack of political solidarity and frequent engagement in out-group favoritism for males, for example, Simone de Beauvoir's existential phenomenological descriptions in *The Second Sex* of women's "bad faith."[2]

Recent debate has taken issue with the theory's oppositional understanding of system versus self, in which system justification is posited as a separate and competing force from self and group justification in the psyches of those who participate in their own subordination.[3] Social identity theorists seek to incorporate an understanding of system-justifying activities as working *in concert* with psychological motivations for positive self-identity formation. Here it may be useful to explore the topic through a return to existential phenomenology, in which questions of the self and formation of identity are of central consideration.

With the aim of contributing to the ongoing investigation into how the seemingly contradictory process of simultaneous self-fortification and denigration occurs, this paper offers a phenomenological study of barriers to feminist consciousness. I examine women's response of *victim blaming* to certain triggering experiences as a psychological mechanism of anxiety abatement.[4] I further explore how the same triggering experiences may instead yield a positive psychological reaction, resulting in the achievement of feminist political consciousness.[5] The processual nature of phenomenological description (as the identification of perception sequences within consciousness) may prove beneficial by highlighting a particular cognitive moment that offers an opening to consciousness raising, as a cognitive fork in the road between patriarchal system justification and feminist resistance.

I employ Robert Corrington's phenomenology describing a particularly ruthless form of system justification—religious violence as justified through fundamentalism. Corrington theorizes religious violence (how acts of extreme brutality are rendered righteous in the minds of community members) through Wilhelm Reich's psychoanalytic concepts of the "emotional plague" and "character armoring."[6] Corrington is interested not only in the violence carried out by one culture against another "alien" group, but when religious violence is directed against one part of a society/tribe against another part of that same society/tribe. Reich's psychoanalysis offers a way of examining such violence as a double infliction of harm: one that is both inwardly masochistic and outwardly sadistic—a turmoil at the individual and communal levels, in which foes are not merely encountered, but created, in order to fulfill certain psychological impulses.[7] Corrington develops this analysis into a descriptive, "ordinal" phenomenological framework that may be used to give an account of the psychosocial processes involved in violent religious movements, such as genocidal attempts to "purify" society, the beheadings of nonbelievers, and sexual slavery of women.[8]

FEMINISM AND THE ABYSS: ARMORING AND DIS-ARMORING

Social and political psychology has extensively considered and studied intergroup violence through a social identity model of "in-group favoritism and out-group derogation." This analyzes how violence is carried out along the lines of in-groups versus out-groups, for example, community members versus foreigners, chosen people versus nonbelievers, and so on. This model is not useful for our purposes, however, since it is primarily suited to examining violence as it occurs between social groups of men and outsider groups of men.[9] The sociopsychological situation of women deviates significantly from this model, in that women are uniquely positioned as a derogated group who are outsiders within their own social structures—in even the most intimate of connections, for example, the sexual and familial.[10]

Women are positioned as a coherently subordinated group, but in isolation from each other physically, socially, and psychologically. Of interest to feminists is the psychological processes and safeguards that prevent women from seeing themselves as a coherent political group and uniting together in common cause with one another to achieve their shared political interests. The forces preventing women's "class consciousness" cannot be traced back to origins from a few hundred years ago, as Karl Marx did for the proletarian class in his historical account of the rise of capitalism. Women's subjugation by men and complicity in that subjugation traces back much further and entwines with some of the founding religious doctrines of current civilization. It is a subordination so ritualized that could itself be analyzed as religion. This was the approach taken by Mary Daly, who described patriarchy as "*the prevailing religion of the entire planet*" (emphasis in original).[11]

Corrington theorizes an "emotional plague" unleashed in religious violence as related to a specific anxiety which inevitably occurs in humans' phenomenological experience of the world. This experience is catalyzed by an encounter with nothingness he calls a "hole in nature"—a concept for describing existential experiences of "the cracking open of an abyss beneath one's feet."[12] He likens the appearance of these holes to "a maelstrom as sea that sucks everything downward," and "a circle with a receding contour that threatens to abolish the self."[13] Examples of such an experience include a sudden pang of existential angst, the grief of losing someone dear, or the disorientation after being betrayed. The grief example is particularly illustrative of the hole-like quality of this experience, in which the loss of the loved one can create a literally empty space in one's life, which was once filled. Corrington defines holes in nature as appearing when "the self fears the loss of identity and an impending sense of fragmentation."[14] This fear of the loss of stable personal contour is often met with a negative reaction in the form of a violent recoiling; the self jerks sharply away from the hole in nature—a reflex rooted in a "deep-seated fear of nothingness in every form."[15] This counterreaction manifests in an urgent desire to cover over the hole at all costs, by which strict identity-preservation takes priority over all else.

Here Corrington employs Reich's concept of "emotional and characterological armoring" to describe the self/tribe's attempt to repress their anxieties through "oaths that bind" in a desperate search for an ersatz wholeness. The self is surrounded in armoring rings, which offer a stronger identity, but at the cost of cutting off the flow of vital energies. The self is divided into two halves: one of armored purity, another of disdainful rejection. A "demonized Other" is often created to further valorize the armored identity through an unconscious negative transference of the impure qualities of the rejected half of the self by projecting it onto some other person/group. Defined in contrast to the self/tribe, the Other is demonized as inherently dirty, shameful, and damned, which may proceed by identifying them as sexually defiled and randomly promiscuous. This framework thus proves highly applicable to describing the psychological processes women undergo in upholding patriarchal discursive practices, particularly when it involves some form of victim blaming for rationalizing male violence against women and girls as something the victims themselves elicited or otherwise deserved.

Imagine, for example, a woman (let us refer to her as "Jessica") who experiences the "rug being pulled out from under her" one day while reading the news. Jessica reads a local news story about a man who last week opportunistically raped and murdered a young woman while she was walking home from a party at night. Jessica notices that the crime happened not far from where she lives and is abruptly shot through with panic at the thought of how it could have so easily happened to her instead. Jessica is considered to be "white" and "middle class," but the security of that identity threatens to be torn asunder by the realization that she could be subjected to this kind of violence just by virtue of being female. A void opens up beneath her threatening to destroy her sense of self and safety in the unexpected pang of awareness at belonging to the sex-caste trapped within what Andrea Dworkin describes as a barricade of sexual terrorism.[16]

But before the black hole can suck her in, she sharply recoils from it and steels herself against it. She armors her heart by turning the victim into a demonized Other: "That foolish girl must not have been careful," Jessica thinks to herself.

> She must have been drunk. She must have encouraged it. She must have flirted with him. She must have been coming back from one of those trashy clubs. What did she expect would happen? One must be smart! Like me. I'm careful. I'm smart. I only associate with *good* (emphasis mine) men. I only go to *nice jazz bars*. I have a good husband, who will protect me, as long as I stay with him. I'm safe and not a *whore* like she probably was.

Jessica succeeds in staunching the flow of female political solidarity that threatened to pour out of her sudden wound and applies several tourniquets. She reaffirms her place in the patriarchal order and the doctrine that tells her she will be safe, so long as she complies with certain expectations: *It is only the "bad" women who will not be spared*. She splits herself in two: into the pure and the despicable. She represses those strange, unruly desires that do not fall in line with the identity of her being a good wife, good mother, good lover—a valuable, desirable woman. Her inner mastery of herself is reinforced through her outward denigration of demonized Other women: those despicable "feminazis," "lesbians," and "whores." She thus cuts down on the possibility of being confronted with that horrifying void in which there is no beneficent male God, and her fealty (i.e., "binding oaths") to the patriarchy was all for naught: a world in which Eve was never made from Adam's rib to be his helpmeet, but rather is a mere tale conjured up by men to convince women to be subservient without question or complaint—a fiction, an unreality, a hole.

The experience of female subordination is haunted by a unique phenomenological void. Simone de Beauvoir examines this existential aspect of female subjectivity as it is compelled in a male supremacist world. She claims that while other people who are socially subordinated as a coherent group, including Jewish people, the proletariat, and African Americans, will often say "We," women do not.[17] She theorizes this as being due to women's uniquely "dispersed" subject positioning throughout their oppressor group, by which their identity is constructed more firmly in relation to certain men in their lives—fathers, husbands, and coworkers—than to other women. As she famously states in *The Second Sex*, "If [women] belong to the bourgeoisie, they feel solidarity with men of that class, not with proletarian women; if they are white, their allegiance is to white men, not to African American women."[18] de Beauvoir theorizes the woman living under male supremacy as the ultimate Other—the subject paradoxically defined as an object—an inessential relation of use value to the absolute, truly sovereign subject, Man.

Following de Beauvoir, women would therefore likely experience confrontations with a sense of female "We"-ness as a loss of a sense of personal identity (i.e., a loss of their identification more firmly with the males in their social groups) and control (i.e., a loss of a sense of power and control over their own destiny gained through their relations to those men). Experiencing such an abyss threatening one's sense of self and stability would understandably be met with horror, such as when the bourgeois woman realizes that her patriarchal allegiance does not stop her own husband from battering and raping her—or that her disdain for those women "lower" than her will not exempt her from the mundane categorical denigration to which all of them are subjected, for example, being viewed as more "emotional" and less "rational" than male peers, or as a permissible target of sexual objectification.

Christine de Pizan's *The Book of the City of Ladies*, first published in 1405, offers the classic feminist depiction of this experience of crisis in women's consciousness as social/political subjects. Christine, the book's main character and self-insertion of the author, is a noble woman with a benevolent father who allows her to spend her time reading whatever she likes, studying science, history, and literature. But one day in her otherwise happy existence, Christine experiences a pang of awareness of belonging to the second sex. She begins reading a book that portrays women as wretched creatures who make men's lives miserable. She puts the book down, thinking it is of no authority, but then is struck by an extraordinary thought. She wonders why it is that the vast majority of men of learning "write such awful, damning things about women and their ways." It is as if all the men in her books "speak with one voice" unanimous in claiming the "vileness" of females, possessing a nature that is "wholly given up to vice."[19] Christine's individual identity and personal pursuits threaten to all be swallowed up in the realization of this one fact, which renders everything she does, and indeed her entire person, categorically invalid on a fundamental level. Embracing the bleak expanse of the "We" of womankind would thus mean accepting that she was forever doomed to being a sinner and a fool, no matter how educated she became, and no matter how virtuous a path she attempted to walk.

Women often recoil from the appearance of the void of wretched female commonality with explanations of their exceptionality. Christine stands at a crossroad where she could respond to this visible flash of abyss by covering it over with a new, stronger identity. Previously, for example, she simply considered herself to be a woman who enjoyed and had made a habit of studying, but she could reconceive herself as instead being *not like Other women* and as more like a man than woman. Christine could further create some demonized Other category of women—women who truly *were* the inheritors of Eve's sinful nature, and deserved, no, *required*, the patriarchal dominion of fathers and husbands.

But Christine does not do this. Her trajectory instead illustrates what Corrington describes as the positive psychological reaction to holes in nature.

Corrington's framework for analyzing intra-tribal violence intentionally leaves open the possibility for religious and personal healing to occur. His use of Reichian psychoanalysis appears to be in part motivated by Reich's rejection of Freud's conception of the death drive as a primal instinct. Reich conceives of aggressive, negative energies as the result of armoring rather than an innate piece of human nature and inescapable compulsion.[20] Corrington's phenomenological description of the yawning gaps in nature, pulling downward on the self, present not simply a malevolent-seeming catalyst for self-armoring, but also an opportunity for the opposite reaction—for restrictive armor to be loosened and even dispelled. "False dualisms" of inner and outer, the pure and despicable self, can be overcome, as well as the negative transference of those abject qualities onto the demonized Other through an emotional plague. The emotional plague, by Reich's definition, resists reason, presenting a set of predetermined conclusions about the world, interrogation into which is usually forbidden. Corrington theorizes the encounter with the hole in nature, although horrifying, as having an ability to "shake" the self/tribe's horizon of meaning in a way that may open it to new insight and even personal growth. This framework of analysis is applicable in describing the psychological processes at work in the difficult and often painful experience of achieving feminist consciousness.

Women experience the disruption of their secure, male-identified horizon of meaning through an intrusion of the "We" of identifying with the whole of female-kind. Christine, at this crossroad, does not recoil from the void in nature, but rather lets it swallow her whole. She does not make herself an exception, but goes down with the rest of her sex, even though it brings her despair. She is overcome with sadness and "disgust," then begins to despise herself and the "whole of [her] sex as an aberration of nature."[21]

Christine is completely despondent. But at the bottom of this dark hole, she finds something. A goddess. Christine is emboldened by the light of "Lady Reason," the first of three divine ladies to appear before her, to elevate her thinking to the "realm of abstract ideas," to dare challenge the male consensus on natural female inferiority. Because Christine does not reactively psychologically separate herself from the rest of her sex, she is able to draw on her own life experiences, along with examples of great women from history, religion, and myth, for counterevidence against such claims. Out of these counterexamples of women with noble characters, she builds the City of Ladies as a testament to the virtue of womankind—a testament that will stand for all time, not as an actual material city, which eventually crumble and fall, but as the rectifying ideas presented in the book itself.

CONSCIOUSNESS RAISING

Corrington also employs a Goddess archetype, which he uses to represent the transformative potencies found in the "depth dimension" of holes in nature. He describes the "Great Mother" as goddess archetype characterized not purely by light and goodness but having a "dual nature" of being both healing and terrifying, as "an ongoing transit between chaos and creativity."[22] It is indeed both healing and "agony," as Dworkin describes it, to gain feminist consciousness.[23] Many women, unfortunately, never pass through the valley of terror in removing armoring rings of exceptionality and Othering long enough to reach the oasis of feminist insight awaiting them on the other side.

Although Corrington offers an interesting and philosophically eclectic framework for phenomenologically describing both the negative and positive psychological reactions to holes in nature, he offers few clues as to how the desirable reaction might be cultivated/encouraged in individuals and wider communities over the undesirable one. It indeed remains mostly a mystery as to why some women react to the same experiential stimuli by forming feminist consciousness, while others only encase themselves in thicker and thicker armor against it and intensify their practices of victim blaming their fellow females. The wide-ranging history of the women's liberation movement has demonstrated that feminists come from an endless variety of social positions and personal experiences, thus offering no evidence of there being any identifiable characteristics, experiences, or sociological patterns that facilitate the formation of feminist consciousness.

Corrington does, however, following John Dewey, emphasize the importance of liberal democratic values of social discourse, which allow for a pluralism of interpretations, ideas, and beliefs to be freely thought and expressed.[24] He discusses the idea that "textual literalism" and religious fundamentalism allowing only a singular interpretation of the "one true book" are key causes of "cultural barbarism" that manifests in the emotional plagues of religious violence.[25]

The creative growth potentialities in encountering forms of nothingness can thus be encouraged in a democratic society that values the pursuit of truth, in which "novel interpretants" are not forbidden, but welcomed. Today, feminist ideas remain forbidden in various ways in much of the world through strict adherence to patriarchal doctrine—if not enacted formally in laws, then informally in social norms and attitudes. Even in the so-called "enlightened" Western world, feminism continues to be reinvented as taboo, with feminists continually denigrated as "hateful," "feminazis," "carceral," "sex-negative," "prudes," and so on, and so forth. A society that promotes values of toleration for heretics of various kinds, however, might carry over into allowing greater toleration for blasphemous feminism, as well, which in turn would allow women greater psychological leeway in their personal experiences to arrive at novel experiential outcomes.

CONCLUSION

In this chapter, I have provided a brief phenomenological study conceptualizing *how* the psychological processes behind both victim blaming and the formation of feminist consciousness might occur in the context of a male-supremacist culture. Further study is needed to understand the contributing factors and reasons behind *why* one of these reactions will occur over the other and how to encourage/cultivate the desired one.

This chapter has also provided insights for social identity and system justification theory, regarding women's participation in their own subordination. The "woman question" in these fields appears baffling: Why would women ever choose to promote a system of male supremacy at the expense of their positive group justification (the female sex) and by extension their ego justification (as a member of that sex)? Previous literature in these fields fails to recognize the psychological extent to which women avoid group identification with the whole of their sex, in order to preserve their sense of self-worth, personal dignity, safety, and security. This paper has thus articulated how women's system justification of male supremacy could work in concert with their ego justification through a psychological process of group disidentification via personal exceptionalism.

NOTES

1. John T. Jost, Mahzarin R. Banaji, and Brian A. Nosek, "A Decade of System Justification Theory: Accumulated Evidence of Conscious and Unconscious Bolstering of the Status Quo," *Political Psychology* 25, no. 6 (2004): 881–919. Jost, John T, *A Theory of System Justification* (Cambridge MA: Harvard University Press, 2020).

2. John T. Jost and Orsolya Hunyady, "The Psychology of System Justification and the Palliative Function of Ideology," *European Review of Social Psychology* 13 (2003): 111–153, 117.

3. Chuma K. Owuamalam, Mark Rubin, and Russel Spears, "Revisiting 25 Years of System Motivation Explanation for System Justification from the Perspective of Social Identity Model of System Attitudes," *British Journal of Social Psychology* 58, no. 2 (2019): 362–381.

4. This is similar to investigations into the "palliative function" of system justification. See Jost et al., "A Decade of System Justification Theory"; Jost and Hunyady, "The Psychology of System Justification and the Palliative Function of Ideology."

5. I employ the definition of "feminist consciousness" provided by Gerda Lerner. She writes, "I define feminist consciousness as the awareness of women that they

belong to a subordinate group; that they have suffered wrongs as a group; that their condition of subordination is not natural, but is societally determined; that they must join other women to remedy these wrongs; and finally, that they must and can provide an alternate vision of societal organization in which women as well as men will enjoy autonomy and self-determination." Gerda Lerner, *The Creation of Feminist Consciousness: From the Middle Ages to Eighteen Seventy* (New York: Oxford University Press, 1993), 14.

6. Robert S. Corrington, *Nature and Nothingness: An Essay in Ordinal Phenomenology* (New York: Lexington Books, 2017).

7. Ibid., chapter 1.

8. Ibid., 3.

9. As acknowledged in Richard Wrangham's "Killer Species," *Daedalus* 133, no. 4 (2004): 25–33.

10. System justification theory also identifies social identity theory's focus on mainly intergroup and interpersonal conflict as a limitation of the approach, Jost and Hunyady, "The Psychology of System Justification and the Palliative Function of Ideology," 115. Our present method is suited to our sex-based inquiry due to Corrington's interest in intra-tribal violence, for example, movements to purify a tribe from within.

11. Mary Daly, *Gyn/Ecology: The Metaethics of Radical Feminism* (Boston: Beacon Press, 1978), 39.

12. Corrington, *Nature and Nothingness*, 8.

13. Ibid., 12.

14. Ibid., x.

15. Ibid., 9.

16. Andrea Dworkin, *Right-Wing Women* (New York: Perigree Printing, 1983), 196, 211.

17. Simone de Beauvoir, *The Second Sex* (Lowe and Brydone: London), 1956.

18. Ibid., 18.

19. Christine de Pizan, *The Book of the City of Ladies*, translated and with an introduction and notes by Rosalind Brown-Grant (New York: Penguin Books, 1999), 6.

20. Corrington, *Nature and Nothingness*. Robert S. Corrington. *Wilhelm Reich: Psychoanalyst and Radical Naturalist* (New York: Farrar, Straus and Giroux, 2003).

21. Pizan, *The Book of the City of Ladies*, 7.

22. Corrington, *Nature and Nothingness*, 120.

23. Andrea Dworkin, *Our Blood: Prophesies and Discourses on Sexual Politics* (New York: Perigree Books, 1976), 78.

24. Corrington, *Nature and Nothingness*. Robert S. Corrington. *A Semiotic Theory of Theology and Philosophy* (New York: Cambridge University Press, 2000).

25. Corrington, *Nature and Nothingness*.

REFERENCES

Corrington, Robert, S. *A Semiotic Theory of Theology and Philosophy*. New York: Cambridge University Press, 2000.

———. *Nature and Nothingness: An Essay in Ordinal Phenomenology*. New York: Lexington Books, 2017.

———. *Wilhelm Reich: Psychoanalyst and Radical Naturalist*. New York: Farrar, Straus and Giroux, 2003.

Daly, Mary. *Gyn/Ecology: The Metaethics of Radical Feminism*. Boston: Beacon Press, 1978.

de Beauvoir, Simone. *The Second Sex*. Lowe and Brydone: London, 1956.

de Pizan, Christine. *The Book of the City of Ladies*, translated and with an introduction and notes by Rosalind Brown-Grant. New York: Penguin Books, 1999.

Dwokin, Andrea. *Our Blood: Prophesies and Discourses on Sexual Politics*. New York: Perigree Books, 1976.

———. *Right-wing Women*. New York: Perigree Printing, 1983.

Jost, John T. *A Theory of System Justification*. Cambridge MA: Harvard University Press, 2020.

Jost, John T., Banaji, Mahzarin R., and Nosek, Brian A. "A Decade of System Justification Theory: Accumulated Evidence of Conscious and Unconscious Bolstering of the Status Quo," *Political Psychology* 25, no. 6 (2004): 881–919.

Jost, John T., and Hunyady, Orsolya. "The Psychology of System Justification and the Palliative Function of Ideology," *European Review of Social Psychology* 13 (2003): 111–153.

Lerner, Gerda. *The Creation of Feminist Consciousness: From the Middle Ages to Eighteen Seventy*. New York: Oxford University Press, 1993.

Owuamalam, Chuma K., Rubin, Mark, and Spears, Russel. "Revisiting 25 Years of System Motivation Explanation for System Justification from the Perspective of Social Identity Model of System Attitudes," *British Journal of Social Psychology* 58, no. 2 (2019): 362–381.

Owuamalam, Chuma K., Rubin, Mark, and Spears, Russel. "Is a System Motive Really Necessary to Explain the System Justification Effect? A Response to Jost (2019) and Jost, Badaan, Goudarzi, Hoffarth, and Mogami (2019)," *British Journal of Social Psychology* 58, no. 2 (2019): 393–409.

Wrangham, Richard. "Killer Species," *Daedalus* 133, no. 4 (2004): 25–33.

Part III

ECOLOGICAL WORLD HORIZONS

COMPARATIVE PHILOSOPHY AND RELATIONAL RESPONDING

Chapter 7

Recapturing World-Loyalty

A Relational Response to Ecological Violence

Katelynn E. Carver

A WARNING: THE STORY THAT FOLLOWS DOES NOT HAVE A HAPPY ENDING YET

Whenever considering the perpetration of violence, in any context, I am always primarily concerned with the *why* of that violence so as to better seek out the *how* of ameliorating it, and ideally—eventually—working to reverse it. Here, I would further contextualize that, when I consider these things, the driving framework is always oriented toward applicability: a conversation such as this can easily, and fruitfully, be engaged and debated among scholars; activists can fight battles; politicians can promote agendas—but where does it translate, how can it manifest in the daily lives of people outside of those contexts? This is the key motivation.

As such, when involved with this engaging and debating, I am always aiming to identify where the ideas can be applied en masse, though more specifically, I am concerned where these ideas land in the sense of the Whiteheadian aeroplane:[1] How does a crop duster come back down to earth in the communities where conversations on violence are often scarce? To that end, I am forever drawn to two groups of people. I think of the community where I grew up: a rural Midwestern township too small for its own post office, though with enough farms to hold annual drive-your-tractor-to-school days at the local high school. I remember getting lost, metaphorically as well as properly *lost*, in the endless rows of cornstalks; I remember planting vast gardens, and tending massive rows of crops and watching stubborn cattle mill about on my grandmother's farm, and her father's—this way of life, this group of people, is what I think of first.

Second, as most people do, but for me perhaps—implicitly or otherwise—embodying that first group's swift and ever-hastening decline: I think of family members younger than myself, for whom I experience a sense of responsibility when I think about violence, and the future, and where we're headed as a human community. Invariably, I also think of the generations to come––the countless birth announcements that flood my social media accounts are proof enough that a declining birth rate in the United States is certainly not pure stagnation—and what they represent, not just in what they will be left with, but also in how they are being taught—or more accurately, in many places *failing* to be taught—to meet with what will be their cosmic inheritance.

What are we leaving them, but also, what will they *do*?

It is a valid question to pose: for many in recent generations, the Earth itself is often quintessentially Othered—a stranger who is not invited into the constructive relationship, kept not just at arm's length but often rejected either consciously or unconsciously, but either way with the same result of neglect at best, and destruction at worst, and neglect is, in itself, a form of violence on its own. While the conversation has shifted to greater awareness and care in recent years, said shift has had the greatest visible significance for Gen Z and mid- to late millennials, who have begun to recognize the relationship between self and Earth by recognizing that relationships as almost irreparably damaged. This recognition, while crucial and undoubtedly one small but positive step toward reparative response, often takes the shape of an unfathomable concern piled atop a mountain of unfathomable concerns—the economy, geopolitics, the threat of war, human rights violations, the resurgence of disease in the face of anti-vaccination movements, to name only a few—that these generational blocks feel so overwhelmed by, that it is often less a sense of apathy than one of self-preservation to shut down in the face of immensity: the modern-day robber barons of the billionaire CEO-class loom as a Goliath without a David in sight. Psychologically, we know that human beings often disengage in the face of too much input or stimulation, and faced here with an inheritance of ruin that these generations did not themselves knowingly perpetrate (often seeing themselves, accurately or otherwise, as victims of the inaction and greed of their forebears and, without the tangible successes of those forebears, left bereft of the scale of resources needed to combat or reverse the damage), to distance oneself from the problem sometimes is simply the biological imperative of our minds. Add to this the oft-requisite environment of being plugged-in and attuned to a twenty-four-hour new cycle of even more stimulation—plus the fact that technology is often the means in which we *un*plug via mindless Twitter-feed scrolling and tapping out "likes"—and the ultimate outcome of distance from the natural world become the rule as a matter of course, and escaping it is only a temporary luxury. Encountering nature in fits and bursts does not allow its needs and changes to become familiar, and therefore a meaningful relationship *with* nature is cut off at the knees from the start. For Gen Z and younger, the aforementioned scrolling and disengagement in the face of immensity was well established as the rule and took (and still takes) the place of learning to ride a bike, or even getting licensed to drive a car: things that in themselves are *not* nature, but at least *pass* nature on the way. With even these being in many ways alien, and largely undesired, the connection to a more personalized communion with nature, one that has a greater likelihood to encourage appreciation and preservation, is left unfulfilled.

Moreover, as has been established in the discussion earlier that faced with such an Other toward whom indifference is the best (and at the individual level, much more common) course of interaction, and violence the worst, I propose two routes to circumvent the destructive Othering here taking place. The first is to make the Other familiar: to engender affection and in turn a desire for the Other's well-being, much what René Girard talks about as the positive aspect of mimetic desire.[2] The second is to spark action upon this newfound familiarity. The second is more crucial, admittedly, but while we're on a tight time frame, we must start somewhere. Thus, baby steps: we begin here with the first concern.

In order to discern how we might engender familiarity and ultimately affection for the Othered Earth, we first must characterize the violence of Othering that has led to the current state of affairs. René Girard's theory of mimetic violence suits the global frame for giving context to the climate crisis, specifically at the level of corporate interests, such as fossil fuels, business conglomerates, and the like.[3] Girard's proposal pervades vast sectors of human activity, through which violence is generated within the context of escalating rivalries based on modeled-and-then-mirrored desire, rendering the particular and unique forms of violence seen in humankind as something that is not originary—and therefore is functionally differentiated from animalistic rivalries and violence—but is instead a by-product of mimetic patterns of human behavior. In other words, a kind of violence that is perpetuated here *against* nature that nature itself has no primary causal hand, and likewise has no conscious defense to mount in response. The individual self, while miming the Other, seeks to achieve some victory, some possession or supremacy in exceeding the Other, by obtaining something that both the self and the Other desire: largely resources in the context at hand, not just their exaction but also dominion over them to be used at odds with their natural means of occurrence.

Both philosophical and empirical attestations across myriad contexts have indicated the enduring power of such violence and its brutalizing consequences for interconnected relationships—and interconnected relationships, from metaphysical, theoretical, neuroscientific, and experiential perspectives, to name just a few, are consistently affirmed as defining much of our human existence.

For Josiah Royce, the necessarily relational community is indeed the cornerstone upon which the unique individual is predicated: "My life means nothing, either theoretically or practically, unless I am a member of a community."[4] The Roycean individual is rooted in the context of society and social relationships, with individuality growing as a development *from* those social contexts as a rule: the individual, then, as a small-scale relational factor *is* significant. However, the community is a "social product," with true community being "created by the hard work of free, self-conscious, self-committed, self-creative moral individuals"—the individual is only significant insofar as it engenders true community.[5] However, with the existence of the individual within a given community, and with the relational element of that community taken as a given, the tendency to actively "other" one's fellow humans and label them as enemies—many times within the context of Girardian scapegoating, though also less fully realized in the sense of sacred sacrifice and instead steeped more specifically in the patriarchal mimesis toward the desired object of power—is pervasive, and the specific instantiation of relationship with the Earth is but one example, albeit perhaps the *most* pressing example there is, given that the community of humanity depends on the cornerstone of the planet upon which it lives, not merely an affiliation or loyalty among many that comprise an individual human but *the* loyalty that makes all other loyalties possible.

When loyalties are divided, however, in such a way that the symbiosis of life on Earth becomes overwhelmingly parasitic, lines are drawn, Othering is enacted, and violence usually precipitates, specifically Girardian violence as treated with herein. For the sake of the present context, what Girard terms the object—the thing desired—is equated with the broadest-articulated end for the sake of the relationship of the Anthropocene to nature: that being power over nature, either explicitly or by proxy. The acquisition of a thing desired by another demonstrates supremacy over the Other. The having of something valued and treasured and wanted deeply by another enacts a power of persuasion, of an even deeper sense of modeling desire itself: you have attained the treasured thing—you must have a capacity, an ability, an enticement that we lack, and so we will engage violently to obtain it. In looking at the catchphrase-worthy "1%" of modern society, this in many ways manifests as a cult of celebrity, of personality, in which we see the success of another, crave it, and both adulate and emulate it until the possibility of actually mimicking it to the point of matching it—or preferably, usurping those who initially modeled it—becomes a viable potentiality: think building the largest hotel, or the creation of a human-made island, or private space flight ventures, or even Calico and Unity Biotechnology, the former Alphabet (Google) company[6] and the latter backed by Amazon's Jeff Bezos and investor Peter Thiel of Facebook fame,[7] both taking aim at overcoming perhaps the *most* natural of processes we experience as part of our communion with the Earth: aging and death.

Here, we begin to see the nature of this pervasive desire for power as a coercive power, a violent power, and it is in this struggle that mimetic desire breeds mimetic rivalry, that in turn breeds violence, because within this dynamic the desired power is not ever actually attainable within the continual feedback loop of mimetic behavior: the self imitates the Other, which imitates the self in an ever-escalating pattern of copying—which, as anyone who has ever xeroxed anything knows intimately: the copy is never exact. The mirror image is always flipped. The photograph is an imperfect approximation of three dimensions condensed into two. The mimetic process is doomed from the start, and, moreover, in *failing* to imitate (dominate) the Other, we fall prey not only to the humiliation of our inferiority—which can provoke retaliatory violence in the name of vengeance or "justice"—but also to the intensification of our desire to mimic the reinforced superiority of the Other. Alternatively, to hypothetically *succeed* in miming the Other, and thus attain what is coveted by them (or to successfully model what is desired *within* them), is to merely reverse and perpetuate the same violent cycle. Here, though, where there is not a unique individual responding to the threat emotively, the concert of processes that comprises nature is reacting to human-created carving out of landscapes, burning down of habitats, and heating up of biomes. Humankind tends to respond to these natural attempts to recuperate as a challenge, however, and simply continues to up the ante. In this way, the process is not only doomed to fail—in the sense that all mimetic violence is doomed in the reality of perfect replication being a myth—but, when there is only one side of the equation actively seeking to create a rivalry (humanity) rather than simply assert self-stability (Earth), it is doomed to end when the resources for either goal are expended: which in this case will be the resources for the survival of society and potentially human life itself are no longer available, and thus the mimetic exchange ends in catastrophic failure—arguably the best exemplar possible for the precarity of mimetic violence at large.

From here, however, we can see a glimmer of hope persisting: the fact that one *cannot* perfectly replicate another, that mimesis *cannot* be achieved without deviation from the model, without at least the slightest integration of contextual novelty, is just *not* a bad thing, it is a *great* thing. This is because the inability to perfectly replicate, in itself, necessitates some differentiation, some newness, which in turn definitionally precludes the devolution into habit or stagnation: Royce's individual-in-society and society-breeding individual thus at play, as "individuals without community are without substance, while communities without individuals are blind."[8] If substance can overtake apathy and disengagement, and vision can overtake out-grouping, we begin to see our way out of the mimetic loop entirely.

Returning now to the Roycean true community being and the creation of that community by unique individuals, such a view attenuates just how effective or worthwhile individual contributions can be: if you sort your recycling and buy “naked” toiletries, is it of any use against the thousands who don’t for every individual that does? On the individual level, likely not: a single person’s recycling will not save them from the consequences of climate change. But on the societal level, a single person has the capacity to impact and inspire their affiliated society, as well as to connect with other individuals within a society of recyclers loyal to the same goal, and en masse, these individuals—in impacting the communities they are loyal toward—can, in fact, be effective.

Further, this is reflective of Royce’s own sense of loyalty as being a moral commitment, with the self-enacting that commitment not blindly out of obedience to a collective, but willfully and with intention for the benefit of a larger community.[9] Causes such as those that require this sort of loyalty are manifested within the shape of a collective group as a rule in order to have effect: a single person cannot change the climate, but many acting in concert can, and the more that do, with growing action and loyalty to the cause, the more impact there is to be seen.

But what of the various industrial complexes and the ills of late-stage capitalism? What of the multibillion-dollar companies polluting with their manufacturing or the air travel that has become commonplace doing such immense damage with a single transatlantic flight? Even millions of loyal contributors to the cause of making the Earth familiar and un-Othered face a near-impossible task against the rich and powerful corporations who enjoy their own seemingly endless loyalty to profit. How does one loyalty outweigh another stood against it, particularly when that other loyalty has far greater clout? This, Royce speaks to easily: genuine loyalty serves the good of the world, while predatory loyalty serves the self. He writes,

> A cause is good, not only for me, but for mankind, in so far, as it is essentially a “loyalty to loyalty,” that is, an aid and a furtherance of loyalty in my fellows. It is an evil cause in so far as, despite the loyalty that it arouses in me, it is destructive of loyalty in the world of my fellows.[10]

But we’ve gone too far, made the Earth too much of an Other, or else, we’re only recognizing all of this too late—so goes the oft-heard mantra of both Earth-familiars and their opposites. What if all this is a lost cause, even unto the decimation of the human species? These lost causes, Royce points out, are the manifestation of our highest ideals. While their scope lies beyond the reach and purview of a single generation, they are nonetheless critically necessary for our moral and ethical lives, both at the individual and societal levels.[11] They shape the mores of our communities, and in fact, the lack of loyalty to the Earth and its needs had likely aimed our society onto the trajectory it tracks today. Realigning with the "lost cause" of Earth-knowing and familiarity is the first step not toward ultimate failure, but toward eventual triumphs that are not merely incidental, but encoded in our cultural sensibilities.[12] They will need defending, redefining, and furthering over time, no doubt, but the lost cause need not remain lost forever.

Having addressed the issue of how we've made nature unfamiliar and how we can overcome that too at least net-neutral indifference to nature, we need to envision how we move toward familiarity in order to engender genuine affection. Across the past ten years or more, publications raising the question of the increasing disconnect of young people to their natural surroundings have been numerous. From articles in such publications as *The Guardian*[13] to academic studies such as the *Organ Environ*,[14] discussion regarding whether child development is adversely impacted by their exposure, or lack thereof, to nature, coupled with such terms as Richard Louv's "nature deficit disorder" causing behavioral problems and serving as a "description of the human costs of alienation from the natural world," all underscore the frank claim that "if children lose contact with nature, they won't fight for it."[15]

Helpfully—and we do so desperately need the help—ecstatic naturalism is well positioned to provide the footholds for a response to this state of affairs. Whereas somewhat similar approaches, such as that of process theology (though it does well to differentiate this meaningfully from some brands of process naturalism), require a degree of classical divinity in at least introducing a god-figure, which either needs to be "sold" to an increasingly less religiously traditionalist populations of millennials or requires such attenuation and contextualization as to lose interest for the masses on the way, ecstatic naturalism makes the argument in one fell swoop: from nature's own "sacred folds,"[16] the potential gifted to the universe via nature's very functions and cycles (*nature naturing*) creates, again, a nearly Levinasian instantiation of Othering that axiomatically allows for the integration of relationality in order to live into the possibilities offered by *nature naturing* to help shape *nature natured* as a vast well of possibility, a liminal promise not yet actualized.

While it may seem that nature, in such a purview, is positioned to do well enough on its own, the Anthropocene renders such a state of affairs moot: human beings *do* enter this process, and necessarily so within the larger understanding of ecstatic naturalism; they play a vital role, and thus it is the responsibility of humankind to make certain that the role played is respectful and supportive in relationship rather than destructive and domineering—that we enter the preexisting majesty of the natural order and participate in the "ecstasy" of that extant cycle so as to make the very best of our place and participation within nature regardless of the dimension at hand, and to move in concert in order to strengthen and prove assets to the furthering and growth of the cycle, rather than serve as deterrents to even its natural, most basic way of unfolding.

Here, it seems the key role of creativity in ecstatic naturalism stands as the most natural and promising inroad to reach a new generation: reengaging them with their natural world at the smoothest possible opening of entanglement within the sacred folding by encouraging active participation in contributing positively to enact enfolding anew as active and mindful participants in the natural order itself. If we gauge the current "stage" of the climate conversation as having a temporal foothold at the turn of the millennium, there are indeed vaunted, visible figures for each generational block speaking and acting on a world stage in the name of climate change: from Al Gore's fade from the political stage to tack his name recognition onto climate studies, to Jay Inslee running for president largely with the aim of furthering the climate conversation (which he succeeded in doing, helping to spur climate-focused forums for the 2020 democratic nominees even after he withdrew from the race), to the work of young activist Greta Thunberg in mobilizing the adolescent and teenage demographic and inspiring millions worldwide to act in combatting the climate crisis. These kinds of spokespeople, woven with the natural effects of the twenty-four-hour global media cycle and the instantaneous feedback loop of social media and the internet at large, have done great work in making the environment familiar as a referent: even if a person isn't *in* nature, the conversation is happening in front of them, being shared and reblogged by them, and read or watched by them. Awareness and concern for self are the first points that are being hit—a recent Washington Post-Kaiser Family Foundation Survey revealed that most Americans believe that climate change is a crisis, and most American teens find the reality of climate change frightening—and while immediacy via these sources is lacking, they are not entirely divorced of action: the same poll found that approximately one in four of these frightened teenagers are acting in some way to fight climate change and seek to repair the Earth.[17]

While not affectively ideal, the fact that news about the environment is incredibly weighted to the negative, and the bulk of the damage done is largely caused by ambiguous mass-market polluters, behind the mask of the "100 companies," lacks specificity.[18] In relation to Corrington, it lacks it-ness: Corrington's elucidation of *nature naturing* as the "nature perennially creating itself out of itself alone" as an alternative to the traditional god, itself lacking concrete singular substance in a way that is both colloquially reminiscent of mystical understandings of divine nature and suited to the "big concept" face-off of climate change—the well-being of the Earth versus everything humanity is doing to diminish that well-being.[19] It is worth speculating as to the efficacy of emphasizing the lack of an "it-ness" in the divine, and that being appealing to a generational group that has arguably witnessed only increasingly violent and destructive models of power in the hands of a vaunted secular parallel to sacrality: economic and/or geopolitical power. When the idea of the sacred is thus shifted to the how, the process, the unfolding that anyone might take part of should they are empowered in measures that do not require the magnitude of the whole but can evolve and complement one another toward effective change.

As such, the comparative egalitarianism of ecstatic naturalism stands as quite possibly a viable, or even potentially straightforwardly appealing means of reigniting a sense of what Paul Stenner terms "world-loyalty."[20] Reinvigorating Alfred North Whitehead's equivalence of the concept to religiosity,[21] Stenner places world-loyalty as opposed to supernaturalism, placing it in line at least tacitly with Corrington's ecstatic naturalism. Stenner, speaking closer to a process view, leans toward theism versus religion, but in the context of ecstatic naturalism, and specifically in focusing on environmental and ecological violence, we can take this identification at face value and see it quite literally: we must cultivate a Roycean loyalty with the Earth as our concern, and its protection or allegiance.

A recent post that came to my attention on social media reinvigorated this specific context not long ago, sticking in my mind and paving a runway upon which to land these musings; said post can be summarized, in terms of its present relevance, as having posed the ecologically oriented question: How can we expect children to rally in defense of a planet they do not *know*? Recent activism has, as evidenced in the prior examples, proven at least somewhat effective, but as discussed earlier, it is also predicated more upon fear in response to violence than upon beauty, more upon the threat of loss and its consequence for the Anthropocene specifically than upon the preservation of and love for the interconnected Other in nature for nature's sake. This tension is worth picking apart in order to determine which approach—or what combination of these approaches—may be most effective to coax humanity (most likely to be seen in younger generations, given the learning curve yet to be determined, but ideally for *all* generations, if we can unlearn toxic attitudes, allegiances, and actions) into positive action.

Here, in moving beyond Ursula Goodenough's distinction between "*beauty* of Nature" and the "*workings* of Nature"[22] the following emerges as particularly illustrative: the aesthetic beauty is much easier to interact with than the complexity of the scientific interconnection of the natural world and its goings-on, and where the two can find functional parallel between *nature naturing* (the workings) and *natured* (the beauty), it begins to translate the building blocks of ecstatic naturalism into more broadly accessible terms for the masses. Goodenough herself notes that these workings "[invoke] . . . a sense of mystery,"[23] which is at least colloquially antithetical to creating familiarity, particularly if engaged only superficially. Given the pace at which modern society moves, and the tendency to abridge and make instantaneous not merely gratification, but experience on the whole, perhaps it *is* necessary to favor the first approach: to cultivate appreciation of beauty-as-aesthetic, and affection in turn as a deep appreciation of said beauty (albeit often a selfish appreciation, centered on the affect associated with experiencing aesthetic beauty). Much as social psychology as made explicit what has long been understood that helping others is also beneficial to one's own well-being, perhaps it is simply human, as the human exists in the here and now, to Instagram a shoreline in the process of appreciating its significance, or hashtag a #sunset as contemplation of its meaning. Is this ideal? Perhaps not. Nevertheless, it may be a viable and effective starting point.

The other main starting point is fear of what nature can do, the impact it can have, from a human perspective. Terror, though often well-founded when it comes to climate change and the state of the Earth in modern times, takes the other side of Goodenough's distinction, the "*workings* of Nature," and does something intriguing, if also inherently disturbing: it combines the physical workings of nature and contextualizes it, colors it with the antithesis of the beauty; characterizes nature in its ugliness within the gaze of humankind, and takes the bastardized workings of a broken, damaged nature at the hands of human vices as the *Krampus* tale to scaremonger and capitalize (though certainly not without merit or effect) on the fear-based communication style of our times. If we are not quite abler—as *yet*—to appreciate at large *nature naturing* in itself, we can appreciate what it looks like when it has gone awry, and what that means for our mortal selves.

With these two prongs—appreciation for aesthetics and fear of a process gone wrong—of the first approach in mind, the second approach—less in contrast and more as a realization of the aim in a more full-fledged and deeply ingrained (and arguably, more faithful) iteration—of fully recognizing the Other's entanglement, within a relational existence, with not only one's own well-being, but one's *being* in and of itself, is a tall order. It calls for not merely an appreciation, or the kind of awe that strikes and then evaporates into the ether, but instead for a sustained sense of wonder that carries a sense of gravitas that seeps deep into one's interactions upon all planes of being as critically, inescapably, and breathtakingly intertwined. More than this even, it calls for this sustained wonder to be allocated and expressed beyond the confines of just what benefits that wonder can elicit for our own human selves, or our own human use. "Blessed be the tie that binds," Goodenough invites, "It anchors us. We are embedded in the great evolutionary story of planet Earth, the spare, elegant process of mutation and selection and bricolage. And this means that we are anything but alone."[24] While such a poetic way of navigating the universe is indeed a sense of taking-in and becoming-part that speaks to a level of intimacy that could very well stop the cycle of violence in its tracks, it is a perspective that is both actively and passively discouraged by the values of modern Western society. Karl Peters reflects, "The realization that we humans have emerged out of [the] long evolutionary process calls us to live responsibly. It calls us to respect and cherish all human beings, all life forms, our planet, and the entire universe as the results of a sacred creative process. More importantly, it calls us into communion with that process, to participate in it in realizing new good for humans in harmony with other life forms on our planet."[25] Such an understanding, when applied to one's everyday comings and goings, flies in the face of the accepted status quo: in the name of Corrington's ultimate aim of ushering both the conversation and the prime understanding of the sacrality of nature from simple aesthetics into the realm of recognition of the "folds" of the natural world and its all-pervasive infinitude, neither made or unmade and with no piece of existence beyond itself. This, crucially, is not to say that aesthetics or human experience are outside of this all-encompassing framework; they are, in fact, inextricable parts of what ecstatic naturalism concerns itself with—a fold within the natural world at hand. It is in such a way that not only relatively familiar models of traditional religiosity find functional parallels for entrée into conversation toward the "transfiguration" of those deistic models, but also connections in popular culture that may very well prove significant in "convincing" the generations most in need of connection to the natural world and respect for its immanent immensity. Films such as James Cameron's *Avatar* series, or video games such as *Final Fantasy*, employ the natural world as paramount; while these are not direct correlations by any means, they can provide an entrance to conversation upon which one can build a culture that embodies what ecstatic naturalism seeks to advance.

It cannot be expected that there will be a wide reaffirmation of belonging to the Earth and to one another when divisionary sentiments are the order of the day; we cannot expect to singularly change the fact that "[we] no longer read the book of the universe"[26] and move toward a collaborative "participation in [the] symphony"[27] of being when reading books is on the downslide, and interpersonal clashing in cacophony is the rule. Overarchingly, then, the first approach is both easier and arguably more realistic in terms of widespread adoption and practice from where we *currently* stand as a society, if still a shallower means of inching toward the stated goal—a Band-Aid on the larger problem, but perhaps effective enough to allow for progress to begin. The second proposal is much more sustainable, but more difficult to enact and maintain. Both, however—though to differing degrees—require paradigm shifts not only in individual thinking but also in societal mores: a new model toward which to apply the mimetic process of desire and enactment.

Once again, however, this is simply the surface, the wading pool, the first steps, but perhaps this is fitting. Perhaps this is most suited to the plight, and subsequent task, at hand. To become like children is, in the biblical context, the way into heaven; in the context of ecstatic naturalism, then, it is the way into balance with the interplay of *nature natured* and *nature naturing*, the fortification of a mindset that can brave the void and not merely live to tell the tale but to thrive and learn to create within, not so distanced from the model of Catherine Keller's understanding of the face of the deep, and the formless, boundless potentiality that can be cultivated there, given a mindset to the productivity latent and waiting to be given some form of even illusory order for the implementing, then exhortations of enfolding the chaotic abyss to unveil the greater protentional within, the glorious promise of what nature itself provides in its own right and in its interactions with the rest of what is—perhaps applying this same approach to what is here presented, and the potential held within its folds, calls us to await that same revelation of unfolding anew.

And thus with an eye toward a misplaced innocence, with the wonder of a child, we might toddle forward: once again—baby steps. We might splash unwieldily in a wading pool for a time, acclimating. But maybe, just maybe, this is where we start from—an end of one way of being, or seeing (or perhaps of failing to see, or landing myopic to the wrong places for sight), to usher toward a start of another. Perhaps Eliot was right in that, and perhaps in our wading we might begin to recognize the patterns, the oscillations of nature within the deep, and come to reclaim what it is that's always been beside us, *naturing*—not entirely unrelated to nurturing, in the anthropomorphic sense, in the Anthropocene we find ourselves within—a companion, neither fully loving nor fully detrimental, but a constant that we are beholden to, that reacts to us as we react to it. Perhaps, then, as children, we can foster change and redefine the hermeneutic we teach the generations to come: not only *how* to address the world around them, but also to equip them with the necessary tools to reverse what damage can still be mitigated.

Perhaps, as children, we have a chance to reframe the story, and begin to write a happier ending.

NOTES

1. Alfred North Whitehead, *Process and Reality: An Essay in Cosmology*, ed. David Ray Griffin and Donald W. Sherburne (New York, NY: The Free Press, 1985), 5.
2. René Girard, *The Girard Reader* (New York: Crossroad, 2015), 13.
3. Ibid., 12.

4. Josiah Royce, *The Problem of Christianity* (Washington: Catholic University of America Press, 2011), 357.

5. Jacquelyn Ann K. Kegley, *Genuine Individuals and Genuine Communities: A Roycean Public Philosophy* (Vanderbilt University Press, 1997), 36.

6. "Calico," Calico, accessed August 19, 2019, https://www.calicolabs.com/.

7. "Unity Biotechnology," Unity Biotech, accessed August 20, 2019, https://unitybiotechnology.com/.

8. Kegley, *Genuine Individuals*, 37.

9. Josiah Royce, *The Philosophy of Loyalty* (New York: Macmillan, 1908), 44–46, 82.

10. Ibid., 118–119.

11. Ibid., 279–283.

12. Ibid.

13. George Monbiot, "If Children Lose Contact with Nature, They Won't Fight for It," *The Guardian* (Guardian News and Media, November 19, 2012), https://www.theguardian.com/commentisfree/2012/nov/19/children-lose-contact-with-nature.

14. Susan Strife and Liam Downey, "Childhood Development and Access to Nature: A New Direction for Environmental Inequality Research," *Organ Environ* 22.1 (2009): 99–122.

15. "How to Protect Kids from Nature-Deficit Disorder," Greater Good, accessed August 20, 2019, https://greatergood.berkeley.edu/article/item/how_to_protect_kids_from_nature_deficit_disorder.

16. Wildman, Wesley J. "Corrington's Ecstatic Naturalism in Light of the Scientific Study of Religion," *American Journal of Theology & Philosophy* 34.1 (2013): 3–16.

17. "Washington Post-Kaiser Family Foundation Climate Change Survey, July 9-Aug. 5, 2019," The Washington Post (WP Company), accessed August 20, 2019, https://www.washingtonpost.com/context/washington-post-kaiser-family-foundation-climate-change-survey-july-9-aug-5-2019/601ed8ff-a7c6-4839-b57e-3f5eaa8ed09f/.

18. Tess Riley, "Just 100 Companies Responsible for 71% of Global Emissions, Study Says," *The Guardian* (Guardian News and Media, July 10, 2017), https://www.theguardian.com/sustainable-business/2017/jul/10/100-fossil-fuel-companies-investors-responsible-71-global-emissions-cdp-study-climate-change.

19. Robert S. Corrington, "Evolution, Religion, and an Ecstatic Naturalism," *American Journal of Theology and Philosophy* 31.2 (2010): 128.

20. Paul Stenner, "Psychology, religion, and world loyalty," *Journal of Theoretical and Philosophical Psychology* 29.2 (2009): 102–107.

21. Alfred North Whitehead, *Religion in the Making* (New York: Fordham University Press, 1999), 60.

22. Ursula Goodenough, *The Sacred Depths of Nature* (Cambridge: International Society for Science and Religion, 2007), xvi–xvii.

23. Ibid., xx.

24. Ibid., 75.

25. Karl Edward Peters, *Dancing with the Sacred: Evolution, Ecology, and God* (Cambridge: International Society for Science and Religion, 2007), 5.
26. Thomas Berry, *The Great Work: Our Way Into the Future* (New York: Three Rivers Press, 1999), 15.
27. Berry, *The Great Work*, 20.

REFERENCES

Berry, Thomas. *The Great Work: Our Way Into the Future*. New York: Three Rivers Press, 1999.

Corrington, Robert S. "Evolution, Religion, and an Ecstatic Naturalism." *American Journal of Theology and Philosophy* 31.2 (2010): 124–35.

Girard, René. *The Girard Reader*. New York: Crossroad, 2015.

Goodenough, Ursula. *The Sacred Depths of Nature*. Cambridge: International Society for Science and Religion, 2007.

Kegley, Jacquelyn Ann K. *Genuine Individuals and Genuine Communities: A Roycean Public Philosophy*. Nashville: Vanderbilt University Press, 1997.

Monbiot, George. "If Children Lose Contact with Nature, They Won't Fight for It | George Monbiot," *The Guardian*. Guardian News and Media, November 19, 2012. https://www.theguardian.com/commentisfree/2012/nov/19/children-lose-contact-with-nature.

Peters, Karl Edward. *Dancing with the Sacred: Evolution, Ecology, and God*. Cambridge: International Society for Science and Religion, 2007.

Riley, Tess. "Just 100 Companies Responsible for 71% of Global Emissions, Study Says," *The Guardian*. Guardian News and Media, July 10, 2017.

Royce, Josiah. *The Problem of Christianity*. Washington: Catholic University of America Press, 2011.

Stenner, Paul. "Psychology, Religion, and World Loyalty." *Journal of Theoretical and Philosophical Psychology*, 29.2 (2009): 102–107.

Strife, Susan and Liam Downey, "Childhood Development and Access to Nature: A New Direction for Environmental Inequality Research," *Organ Environ* 22.1 (2009): 99–122.

Wesley, J. Wildman. "Corrington's Ecstatic Naturalism in Light of the Scientific Study of Religion." *American Journal of Theology & Philosophy* 34, no. 1 (2013): 3–16.

Whitehead, Alfred North. *Process and Reality: An Essay in Cosmology*. Edited by David Ray Griffin and Donald W. Sherburne. New York, NY: The Free Press, 1985.

———. *Religion in the Making*. New York: Fordham University Press, 1999.

Chapter 8

Fecundity and Healing of the Great Mother

Reading Corrington's Nature and Nothingness *via Yin-Yang Thinking*

Jea Sophia Oh

INTRODUCTION

Both women and nature as the bottom of anthropocentric hierarchy have been subalternized by radical anthropogenic climate changes across global, regional, and local scales. Nonetheless, victimization of the subaltern not only weakens the power of subalternity but also limits its boundary. My comparative reading here of Corrington's *Nature and Nothingness* via the dynamic process principle of yin-yang will seek to provide a healing philosophical framework and analogy to address this time of rampant species endangerment and mass extinction, the global climate emergency, and the increasing rise of racist and xenophobic authoritarianism. I introduce Korean term *salim* (enlivening) as the healing spirit amid our pandemic/pandemonic time now along with Corrington's notion of the Great Mother. These archetypes don't need to be essentialized or genderized as a metaphor for specifically women only. Despite the genderized degradation of both women and *salim*, activities of *salim* make things alive and enliven nature narrowly in a family and broadly on this planet. *Salim* cannot be genderized or essentialized as masculinity or femininity but rather speaks as a transgressive "fecundity." This unleashing power of life exists in nature, in every living and nonliving aspect of nature as the healing energy. Life comes from this hidden energy (yin) which is ceaselessly manifested (yang) in nature through the circles of yin-yang (陰陽).

In *Nature and Nothingness: An Essay in Ordinal Phenomenology* (2015), Robert S. Corrington focuses on the how of nature in a broadly comparative methodology of ecstatic naturalism—there is a strong ethical dimension to this work as he thinks through how the various "holes in nature" can help keep the Selving process free from authoritarian impulses. Corrington unfolds the dynamic relationship between nature and nothingness by conceptually tracing the marvelous the marvelous logarithmic spirals at play in constituting the fecund holes of nature that are on an ontological continuum with other "forms of nothingness" allowing for an adventurous journey into totalizing nothingness, naturing nothingness, and encompassing nothingness. Corrington's spiral of nothingness indeed resembles yin-yang (陰陽) thinking and the always striving processes that *nature naturing* and *nature natured* (*Dao*道) have toward achieving "the Supreme Ultimate" (*Taijí* 太极).

First of all we do well to note that "the holes of nature" can be read analogously to the myriad things of yin-yang, a kind of natural "subaltern" that really suffers most from the ecological crisis and various forms of political repression and hegemonic oppression. Second, we have "totalizing nothingness" which can be interpreted as an extremity of yang or an imbalance of natural and social forces showing the evil of humanity and at the same time revealing nature's suffering under the triumphalist ideologies of authoritarianism. The result of the U.S. presidential election 2016 shows that the neoliberal capitalism of America finally touched a boiling point of its unbalanced extremity of an unbridled patriarchy or a yang (masculinity) without balanced harmony, and this transformative moment predicts or prophesizes the arrival of its alterity, a resurgent power of yin (femininity). Third, we have "naturing nothingness," as a culturally creative emanation from the collective unconscious of *nature naturing* that is predictively invoking the arrival of a powerful resistance and healing yin (femininity) power, which can be effectively seen in then Women's March in Washington, DC, held on January 21, 2017. The march was not by and for women only, but by and for all subalterns including the subaltern nature as our excluded and exploited planetary ecosystem.

Therefore, Women's March was a march for regaining access to the creative fecundity of nature beyond a singular or reductionistic conception of femininity. The entangled bodies as political assembly represented the concretized affects of the people's collective rage against injustice. I understand the Women's March as the subaltern's performative gesture of a spatial and critical intervention into the totalizing power of the hyper-capitalistic and patriarchal America. Such an embodied and collective intervention can be part of a healing process of the broken harmony of yin-yang as the ethical and political reaction to the election outcome. Finally, there is "encompassing nothingness" which can be compared to the Supreme Ultimate (*Taiji* 太極) of yin-yang. Corrington calls nature's encompassing power a god-ing capable of transcending ordinariness by enhancing both personal and social forms of life. This cultural optimization and divinization of life in Corrington's notion of god-ing (a notion that remains skeptical of transcendent monotheism) is similar to the Korean idea of *salim* (enlivening). *Salim* represents the intrinsic healing power of nature and can make even wounded and dead things healed and revived if given the proper nurturance and timing. The radical potential that nature has to enliven goes beyond mundane thinking and is akin to what Corrington expresses as the aim of allowing nature to become "free from its captivity in bad forms of metaphysics and can shine forth as the glowing infinity that it is."[1]

THE HOLES IN NATURE: THE SUBALTERN

> Nature as "wheel of agony," rotates around one spot, dispensing woe to all creatures. . . . Yet nature is dual in its operation. On the one hand it is "ceaseless devouring," while on the other it is "giving birth."
>
> —Robert S. Corrington[2]

> 貴以賤爲本
> The humble is the foundation of the dignified.
> 高以下爲基
> The inferior is the ground of the superior.
> 是以侯王 自謂孤寡不穀
> Kings consider themselves "orphaned," "widowed," and "insignificant."
>
> —*Dao De Jing* 道德經[3]

On January 21, 2017, the day after the inauguration of President Donald Trump, the Women's March in Washington, DC, convened as a women-led movement bringing together people of all genders, ages, races, cultures, political affiliations, and backgrounds to affirm a shared humanity and to pronounce a bold message of resistance to patriarchy, misogyny, racist, and other forms of hatred, and to practice self-determination in political space. The Women's March was a worldwide protest movement advocating legislation and policies that would respect human rights and other issues, including women's rights, immigration reform, health care reform, LGBTQIA+ rights, racial equality, and earth rights. It was the largest single-day demonstration in the history of the United States. It was organized as a movement of the subaltern to send a bold message to the new administration on their first day in office and to the world that "women's rights are human rights as well as the earth rights."[4] The new presidency of the United States forebodingly signaled the dawn of a new kind of reactionary conservative movement in the United States, which, with an ideological foundation of "aspirational fascism" and a political tendency sometimes known as Trumpism, a whole package of bad ideologies converged: racism, homophobia, misogyny, elitism, classism, speciesism, selfish-anthropocentrism, and so on. The U.S. election results invoked an intensification of negative affects like fear, anxiety, hatred, and anger from persons across the ideological spectrum. The Women's March was a positive and hopeful reverberation with all of the postelection trauma. Nonetheless, the march was not by and for women only but by and for all the subalterns including nature as our planetary system. We might ask then as a consequence of neoliberal capitalism, did the march mark an end of progressive liberal multiculturalism and the dawning of a new Dark Age in global politics? Or, on the other hand, might it be a wonderful sign of a parousia of yin (femininity)?

Corrington makes the interesting suggestion that nihilating is being probed by nothingness in bridging the thinking of Martin Heidegger's *Being and Time* (1927) and Jean-Paul Sartre's *Being and Nothingness* (1943). Corrington uses his ordinal phenomenological method to observe the modes of nihilation of nature by delineating four types of nothingness: holes in nature, totalizing nothingness, naturing nothingness, and encompassing nothingness. Corrington calls the "holes in nature," finite and manifest themselves in *nature natured*, which appear whenever the self fears the loss of identity and an impending sense of fragmentation, as the holes appear malevolent in nature.[5] Nonetheless, Corrington's holes are not really static structures. Rather, holes of nature in Corrington are similar to how he conceptualizes an abyss, which he calls "naturing nothingness," because it is not an ontological vacuity but a dynamic force that remains flexible and active in the pulsing emanations of *nature naturing*. The holes of nature are the enabling conditions for any nature natured.[6] They are eventually filled and healed by nature and become ingredients and conditions for new becomings in its perennial nihilating momenta.

Once the holes of nature become the conditions of new becomings, they can yield good and evil, and positive or negative things; although how they can bring positive or negative events and effects depends upon how they are connected with other entities and natural complexes and vice versa. Nonetheless, the traces of holes are always still there as the karmic causes, not in reduced or expended forms, but in always transforming hybridic novelties. Therefore, the holes of nature are inviting and waiting for the next interested parties to engage in transformative spatiotemporal conditions which are created from any separations, hardships, lacks, trauma, violence, death by natural and unnatural causes, and so on.

"The holes in nature" could be a chance for the colonizers to take the life of the colonized. In such cases, the holes of nature are evidence of colonization and harbor the existence of the subaltern and the possibility of reversal and deconstruction, which might eventually call for a reconciling harmony and healing. The Women's March movement represented the interconnected whole of the subaltern, the myriad things calling out as the so-called orphaned, widowed, and worthless, which is how ruling superiors should consider themselves according to *Dao De Jing*, Chapter 39. This chapter of the Daoist philosophical classic teaches the interconnectedness of the myriad things by claiming that the inferior is the root of the superior and the low is the foundation of the high. This chapter is encoded with a dynamic principle of yin-yang (陰陽) as the two vital forces of *ki* (氣). The text presupposes that despite seeming to be two antagonistically opposed or contrary forces, the dipolarity of vital energy is actually complementary, interconnected, and interdependent in the natural world.

This interdependency of nature is Dao. And if a ruler disregards his dependency on the people and does not identify himself as orphaned, widowed, and worthless, he all too easily becomes a dictator, or an aspirational fascist. Nonetheless, "we the people" as a phrase deployed in the Preamble to the U. S. Constitution maintains the freedom of assembly. Women's March as an assembly didn't mean only occupying a space by the multitude, but it was also indicative of a radically pluralizing movement of bodies as the subaltern resisting monopolies of power. Indeed, the bodies of the subaltern are almost always taking the shape of decentralized power/movements, and as Hannah Arendt's famous claim made clear, politics requires not only a space of appearance but also bodies that actually do appear.[7] An assembly creates a space of appearance and gives a new meaning to the place as Arendt writes, "Action and speech create a space between the participants which can find its proper location almost anywhere and anytime."[8] Judith Butler also recognizes the vital importance of the political appearing in the streets:

> In the case of public assemblies, we see quite clearly not only that there is a struggle over what will be public space, but a struggle as well over those basic ways in which we are, as bodies, supported in the world—a struggle against disenfranchisement, effacement, and abandonment.[9]

And for Butler, it is obvious that "the assembly is already speaking before it utters any words, that by coming together it is already speaking an enactment of a popular will; that enactment signifies quite differently from the way a single and unified subject declares its will through a vocalized proposition."[10] These assembled multitudes as the spontaneous and decentralized organization of many bodies in political movements of resistance and contestation are not like the random crowds in Times Square in New York City during the holidays, but they are the assembly of the subaltern bodies, the plural movements, the democratic power of the people—a "we," that can effectively demonstrate the existence of the subaltern and at the same time decentralize, deconstruct, and dismantle the monopolies of hegemonic, capitalist, and patriarchal power by occupying the streets. Such occupation of the political space of appearance enables and enacts a symbiotic body/power of the myriad bodies, as just so many "holes in nature" orchestrating to become a moving whole.

TOTALIZING NOTHINGNESS: THE EXTREMITY OF YANG

> There is something deeply embedded in the Selving process that retains a fascination in horror, whether in literature, film, or other modes of artistic expression. This natural indifference, especially in the innumerable orders of nature natured, shocks the self and reminds it of the absence of teleological patterns in organic nature, not to mention any ultimate salvation within *this* world (emphasis in original).
>
> —Robert S. Corrington[11]

> 民之饑, 以其上食稅之多, 是以饑,
> The people's starvation is because those above take too much taxes.
> 民之難治, 以其上之有爲, 是以難治,
> It is difficult to govern the people because those above manipulate them.
>
> —*Dao De Jing* 道德經[12]

Paradoxically, Corrington explains "totalizing nothingness" via two poles of transcendental powers functioning beyond ordinariness. Nature's sublimity is a greater inclusive aesthetic value than forms of beauty, which typically can only be found in works of art. By contrast, the sublime contains within it forms of beauty but also transcends mere beauty by touching upon the infinite abysses of nature. Perhaps nature's sublimity can be found on the edge of decay and death. However, there is no category or conceptual containers that can exhaustively capture or hold the sublime, for the sublimity of nature is always going beyond the clear spatiotemporal boundaries around the edges of beauty. Corrington says that it is in this sense that "the sublime is totalizing and thus has a direct link to the *totalizing nothingness*" (emphasis in original).[13]

Fundamentally, the Western monotheisms have a totalitarian tendency, a theological tendency which both Corrington and Ursula Goodenough abhor and criticize. Goodenough says, "For me, Deism doesn't work because I find I can only think of a creator in human terms, and the concept of a human-like creator of muons and neutrinos has no meaning for me."[14] She seems to be suggesting here and elsewhere an alternative religion, which takes seriously the mystery of nature on its own terms without introducing a single-source, anthropomorphic creator deity by saying that "mystery generates wonder and wonder generates awe."[15] For similar reasons, Corrington recognizes the superiority of art to these religious pathogens: "Unlike religion, art does not unleash the will-to-power; that is, the drive to overpower rebellious "insiders" and smash the 'idols' of 'outsiders.'"[16] One of the reasons he provides for this superiority is that art is not inherently iconoclastic. In my essay, "Delivering Peace Out of the Broken Womb," I also criticize the entanglement of capitalistic American neocolonialism and the totalitarian aspects of monotheism: "Any religion that practices *jugim* (violence) against life becomes an empire. No religion can claim to contain all truth. It also cannot be denied that there are elements of truth in the different religions."[17] The monotheistic homogenizing patriarchy of Judeo-Christian religions is indeed matricidal and racist and has reinforced other oppressive systems and been internalized in the triumphalist Americans to accept a racist exploitative neoliberal capitalism by bringing a psychic trauma inflicted on the earth and the lower classes.

Another aspect of "totalizing nothingness" can be found in the horrors in human history such as ongoing genocides, endless religious wars, systemic racism, environmental violence, and the like by totalizing power that exercises dictatorial control over many aspects of life, relating to a centralized governance and political sovereignty, and exercising control over the freedom and will of others. In considering the potentially transformative and healing potential of even the horrific (as an aesthetic experience) Corrington states that "to encounter horror is to be shattered by the totalizing nothingness that can engulf the human process."[18] Since the ugly horrors of extreme violence and torture transcend the ordinariness and the capacity of the majority of our species, it will eventually bring resentments and countermovements. We need to imagine differently the political aesthetics that valorize the fetishization of violence, white supremacy, and patriarchal fantasies of total domination.

Dao De Jing, Chapter 75, teaches that a major reason that people are rebellious is because of excessive taxation and heavy-handed interference in the ordinary lives of the people. The repetition of the same deadly and dehumanizing pattern of police brutality against African Americans in the United States has unleashed an immediate and unrelenting national and global uprising for Black Lives, unprecedented in its global solidarity for racial justice. Since Eric Garner's dying word, "I can't breathe," in 2014 was repeated by George Floyd in 2020, the Black Lives Matter movements have become a major cultural phenomenon and a force to be reckoned with in global networks of survival power bringing justice, healing, and freedom to black and otherwise marginalized peoples across the globe. The expansive movements have been building collectivities that believe in an inclusive and spacious movement for justice beyond narrow nationalisms. According to Roger Ames and David Hall's philosophical interpretation of *Dao De Jing*, Chapter 75:

> The people's lack of order is because those above manipulate them—and the people's scoffing at death is because those above are exacting so much from life, the general premise is that much of not most of the suffering of the common people is an immediate result of bad government.[19]

To speak truth to power is not fundamentally an individual act. Judith Butler argues that before we ask what it means to speak truth to power, we have to ask who can speak. Given that a Spivakian analysis of any individual subaltern is not able to speak in front of power because she is involuntarily rendered mute by the hegemonic languages and power structures that systematically effaces her voice. Nonetheless, the power of the subaltern is not just a solitary, poor individual, but is composed of the many bodies of people who are supposed to remain mute in public discourse but choose to instead deconstruct and dismantle the hegemonic power structures by gathering together in the streets. This is the other side of power, the power of survivors as well as the survival power; in Foucauldian terms it is a bio-power, which decentralizes the legitimate power. One has to consider the heterogeneity of power as articulated in Michel Foucault's work. In *The History of Sexuality I*, Foucault views sovereign power as representing itself as negating, legislative, prohibitive, censoring, and homogenous while he recognizes bio-power as always productive, immanent, exercised, capillary, and resisted.[20] Power also comes from the bottom of the structure, which has been considered as the powerless. Thus, power should be interpreted in terms of multiplicity and fluidity.

"Totalizing nothingness" can be interpreted as the extremity of yang showing the evil side of humanity and at the same time revealing nature's suffering. Trumpism(s) are a particularly ugly and virulent offspring of the globalized reactionary movements seeking to maintain racial and patriarchal purity in the face of a globalizing neoliberal capitalism. Part of what a healing movement to such fanaticism requires is a transformative corrective to the primacy given to selfish competition which teaches people to be psychopathic and atomized individuals cold to the suffering of others, and expressing no empathy for the oppressed and subaltern. The need to do psychoanalysis of political economy is what motivated Gilles Deleuze and Félix Guattari to relate capitalism with schizophrenia. Deleuze writes, "In capitalism, desire and interest or desire and reason are distributed in a totally new way, a particularly abnormal way. Capital, or money has reached such a stage of delirium that there would be only one equivalent in psychiatry."[21] For Deleuze, capitalism has always been and still is a remarkable desiring-machine: "Flows of money, flows of the means of production, flows of human-power, flows of new markets."[22] And along with the schizophrenic-inducing capitalism, the extreme racism of the United States is part and partial of the psychiatric disorder of a fascist "characterological armoring" which is seemingly touching a turning point of its extremity in patriarchally unhinged yang (masculinity) that is signaling the coming arrival of the time to address and be addressed by the subaltern, yin (femininity). Thus, the people's rebellion against a bad political order is a natural transformation pivot according to *Dao De Jing*.

NATURING NOTHINGNESS: THE ARRIVAL OF YIN

> The Great Mother functions as the goal of the Selving process. She must be freed of patriarchal distortions. . . . Even if a male is seeking the misconstrued maternal, the Great Mother, the creatrix of nature, lies beyond such projections, fantasies, male longings, and male-driven hierarchies. One must trust the natural grace that comes from her manifestation.
>
> —Robert S. Corrington[23]

> 谷神不死 是謂玄牝
> The valley spirit never dies; It is the black female.
> 玄牝之門 是謂天地根
> The gate of the dark mother; It is the foundation of heavens and earth.
> 綿綿若存 用之不勤
> She continues life like delicate threads; It is bottomless productivity.
>
> —*Dao De Jing*道德經[24]

In fact, people's movements cannot be homogenized into one type of ideological stance or as a monolithic response to systemic problems. Anger and rage stemming from wrongful deaths and killing from the totalizing violence of state and public power become a powerful social emotion that can be transformed into vitalizing people's movements from the micro-level of individual survival to the macro-level of addressing ecological and economic systems—such people's movements as forms of social organization always operate through the complex circuits of power that connect and mobilize these different strata of social realities. Sara Ahmed has emphasized that we have to read carefully whereby pain and painful affects are moved into a public domain, and subsequently transformed into potentially democratic catalysts of change. Certain feminist projects might become a way of responding to the pain of others, and Ahmed asserts that some feminisms even claim to stand for or even stand-in for the original, direct experience of being wronged or being pained.[25] However, any form of feminism has nothing to do with pain, if there is no actual compassion in relationship with actual others.

The Women's March assembly was the objectified "affects"[26] of the people's collective anger against injustice. Suffering can cause both anger and compassion. I recognize the Women's March as an instance of the subaltern's performative gesture of spatial and critical intervention into the totalizing power of the hyper-capitalistic and patriarchal America. The original movement and subsequent developments might herald the coming of a new society and serve as a remedy for the broken harmony of yin-yang as the result of the Trumpist regime. The march was an embodiment of affects, especially the pains of the subaltern, and demonstrated that we are able to resonate with one another's feelings, and no longer had to feel atomized, afraid, and alone. Deleuze and Guattari write that affects are always "becomings."[27] The Women's March was a collective becoming of objectified affects through the various Selving process of resonating with others' sufferings as a collectivity of subalterns. However, if any of the Women's March movement excluded LGBTQIA+, earth rights, Black Lives Matter, and other vital movements for justice then it becomes nothing more than a liberal, white-Women's March. All types of subalternity should be considered and included when assemblies gather in public spaces in order to fight against the decimation of shared infrastructural goods and for the coming presence of true justice. The subaltern will be enhanced in their powers and expanded in the scope through their collective assemblies and living with compassion beyond the dividing and conquering boundaries of race and gender.

Therefore, the feelings of the people can be transformed into political movements and practical ethics. Benedict de Spinoza defined "bondage" as a man lacking the power to moderate his affects.[28] For Spinoza the field of ethics is not marked by a dualistic dichotomy between Reason/Feeling, on the contrary, "the longitude of affects becomes ethics"[29] which we practice precisely by resonating more keenly and intensely with the suffering of others including the right to go out onto the street that is the "will to bio-power," as a Selving process involved in healing collectivities. Compassion is a spiritualized act of affective life as well as the basic element of any functional democracy, which has to focus more on the *demos* (people as the subaltern) than any *cracy* (attempt at patriarchal dominating).

Roger Ames and David Hall translate "*Xuánpìn* 玄牝" of *Dao De Jing*, Chapter 6, as "the dark female (*Xuánpìn* 玄牝),"[30] which is the vaginal beginnings and vital root of the world, a source of bottomless procreativity. Here, the dark, moist, and accommodatingly vacant interior of the vagina is used as an analogy for the ultimate fertility of *nature naturing*—a communicative and transformative portal with no harmful teeth (the mythological antipode to any horrific *vagina dentata*). The primal mother is indeed a symbolic metaphor of fecundity of nature or *Life per se*. In Corrington's lexicon of ecstatic naturalism the dark female is the archetypal abyss of the primal mother. The primal mother inexhaustibly keeps giving birth to the myriad things. However, the primal mother can't be a singular female. They are the archetypal super productive and reproductive fecundity of all that is happening in the unsummed totality that are the innumerable ordinal complexes emanating from the "encompassing nothingness" of *nature naturing*. Compassion is the valley sprit of *Dao De Jing*, and is intimately involved in the anti-fascist Selving project for Corrington, as the emanations from the abyss of the primal mother call us to trust and hope in the restorative and healing powers of nature rather than put up defense mechanisms and hard shells of stultifying characterological armoring. For Corrington the allure of the primal mother as a semiotic fold in nature do not imply a fixed identity or simple location for the "emanations do not come from one primal emanation but are as perennial as nature itself."[31]

I recognize the Women's March, along with other subaltern movements such as the post-*Sewol* candlelight protests in Korea[32] and the global Black Lives Matter movements, to be instances of what I've termed else an "*Ugeumchi* phenomenon" which is what Kim Ji-Ha, a poet and playwright of *Donghak*, used to refer to the movements of the subaltern. The *Ugeumchi* War (1894) was the last and the most desperate battle of the peasant rebellion *Donghak* movement that occurred at the top of *Ugeumchi* mountain pass where a rebellious military force clashed with the established gentry system in order to deconstruct and dismantle a corrupted aristocracy and respond to various modalities of institutionalized injustice. Kim realized the fact that the mystery of life is a synergistic transformation of yin and yang by observing several small fish successfully swimming against a strong current in a polluted stream. The stream had the downstream trajectory of forced yang motion, while the small fish were able to counteract the current to follow the pattern of nature's resisting and returning (*fan* 反) by a receptive and adaptive yin motion. Kim realized that the fish utilized the latent yin energy of the water although the water's trajectory was a flowing downward.[33] Thus, the yin-yang dynamic of the stream and the fish resonated for Kim to create a miracle through the cosmic activity of *ki* (vital force) and served as a powerful analogue for the harmonizing out of political struggle and strife through the continual exchanges of collective diffusion and convergence. The *Ugeumchi* phenomenon was the movement of the subaltern spirit, yin spirit, the divine energy (*shin-ki* 神氣) to regain true subjectivity and a critical consciousness against oppression, even if the general trajectory seemed impossible to overcome. Thus, it signified a movement of the subaltern's "collective conscientization (*conscientização* in Portuguese)"[34] against injustice.

ENCOMPASSING NOTHINGNESS: THE SUPREME ULTIMATE (*TAI-CHI*, 太極)

> Nature becomes free from its captivity in bad forms of metaphysics and can shine forth as the glowing infinity that it is.
>
> —Robert S. Corrington[35]

人法地, 地法天, 天法道, 道法自然.
Humans follow the earth;
The earth follows the heavens;
The heavens follows *path-making*;
The *Dao* follows nature's *self-soing*.

—*Dao De Jing*道德經[36]

The power of subaltern, yin sprit, the Great Mother, doesn't belong to women only. The principle of yin-yang is not a dichotomy of women and men, but a great harmony of the myriad things through the flows of *ki*. Yin-yang should not be misconstrued as two fundamental structures of a dichotomized gender identity with femininity perpetually opposing masculinity, for that would only further solidify global patriarchy and the semiotics of phallogocentrism. Butler avoids any feminist essentialisms by steering clear of the totalizing gesture of feminism. There is no universal femininity, but only so many embodied women. She writes, "The insistence upon the coherence and unity of the category of women has effectively refused the multiplicity of cultural, social, and political intersections in which the concrete array of 'women' are constructed."[37] The Women's March then is best conceived as an interconnected ensemble of bodies rather than as some exclusively female or gender neutral movement for which there can be no unique space of appearance for differently gendered persons, but only a futile resistance against injustice and at best token displays of compassion with sufferings of others.

The Great Mother, yin, is not just archetypal femininity but a caring "fecundity." Beyond the dichotomy of masculinity and femininity, everything is endlessly creating myriad things in the process of pure becoming. In this sense every entity, both at the micro- and macro-levels, can be semiotically depicted as "a maternal body."[38] This deep hidden power exists in nature and works not only through women, men, and transgendered persons, but also in and through every living and nonliving aspect of nature. The asymmetrical relationship between women and men as conceived by the paternalistic and patriarchal conception of "law and order" is precisely the problem. The patriarchal monotheisms and the global capitalistic economy lack a balancing yin principle. So we have to rescue and revive "fecundity" out of this unbearable order.

In this sense, the Women's March heralds a subaltern assembly that is freed from the shackles of the paternalistic law, and may well prove to be yet another incarnation of the fecund yin principle. Therefore, the Women's March is best recognized as a march for regaining the caring fecundity of nature beyond essentializing femininity. Yin-yang is not a dichotomous structure, but a transforming spiral of "the Supreme Ultimate (*Tai-Chi*, 太極)." This way of thinking can offer a hopeful message and a warning for the subaltern, especially for the suffering ecosystem wrought by the totalization of greedy humans: namely, that, even despite all appearances to the contrary, "life returns to the Supreme Ultimate, the Great Harmony." Corrington proposed his fourth form of nothingness as the "encompassing nothingness," a mode that encompasses both *nature naturing* and *nature natured*: "It is the ultimate form of nothingness and nihilation as it is the surround for all of nature."[39]

For Corrington, "nature" does not refer to anything per se, but is itself the dynamic entirety, an extremely wide and deeply vast reality that perpetually creates itself out of itself alone. Perhaps then "encompassing nothingness" is an approximation of *nature naturing* or nature's expansion. There is nothing outside of nature and nature is the ground of the sacred in ecstatic naturalism. Corrington emphasizes that "the sacred is the holy of nature," and as epiphanies, "nature's sacred folds show forth something that is extrahuman, although not extranatural" in its mysteries of nature itself.[40] In his early work, *Ecstatic Naturalism: Signs of the World* (1994), Corrington echoed the famous Greek phrase by replacing God with nature: "Nature itself as an encompassing that has no outer shape or circumference."[41] Hopefully, as Corrington's ultimate sentence makes clear in *Nature and Nothingness* (2017), "Nature becomes free from its captivity in bad forms of metaphysics and can shine forth as the glowing infinity that it is."[42]

CONCLUSION: HEALING EROS OF THE GREAT MOTHER

As I proposed earlier, the fecundity as the ultimate power of life, a power which follows the dynamic principle of yin rather than an unbalanced yang, although yin can still be recognized as femininity in order to overcome dualistic patriarchy, is what Corrington calls the power of nature to become open to the community of interpretation as "the patriarchy-free Great Mother."[43] Corrington asserts that "the most healing of all the archetypes is the Great Mother archetype" who has been largely lost/absent, distorted, even murdered in almost all religions, not just the Western monotheisms.[44]

Can all the entanglements of traumatic wounds and pains from patriarchy /sexism/xenophobia/racism/classism/speciesism/and so on be healed by the erotic and healing power of the Great Mother? Corrington might be naively hoping that the Great Mother archetype, in her gathering and healing Eros, can actually dissolve hierarchies in all modes, be they ecclesial, political, and social, or economic. Nonetheless, having a Mother Goddess such as the Great Mother archetype doesn't necessarily translate into a non-patriarchal society. The Great Mother archetype is a nonessential/anti-stereotypical principle of fecundity and love for life (Eros for Corrington; *Salim* for me).

Eros is the great leveler and is a force to reckon with in Corringtonian ideals of ontological democratization.[45] For Corrington, Eros is the spirit of nature itself. It is *nature naturing* generated by betweenness, the lure of nature like a great openness and unconditional invitation to new organismic entanglements together on the expanding edges of the entanglements. The healing of Eros is found in the unprofitable/noncompetitive/noninstrumental economy of desire so very unlike capitalism. With the healing Eros, we all become more vulnerable to be open to being an intimate part of the planetary and polyamorous expansion of the Great Mother in new entanglements together that might serve to distance individual selves, but tie one another together as *Nature*.

A prominent theologian of earth in our eco-catastrophic time, Catherine Keller in her new book, *Facing Apocalypse: Climate, Democracy, and Other Last Chances* (2021) poses a serious counter-apocalyptic question on our common belief/hope for healing erotic metaphor of cosmopolitan synonyms of peace and love:

> We have never lived elsewhere, anyway. Perhaps a down-to-earth regenesis of our life together is what the new heaven and earth was always about. Or does such improbable *shaom, salaam, salim*—reveal itself as all too utopian? . . . Could we, some critical mass of the species that has trashed the earth, yet recognize it as "cosmic sanctuary"?[46]

"Yes, indeed we could and should!" *Salim* is a Korean term, which means making things alive, restoring, healing, and enlivening. Extensively, fecundity and healing of the Great Mother can be recognized as *salim*. Embracing the other is a way of living and expanding self with others as a transcending whole via our symbiotic care. Nonetheless, it has become more and more difficult to touch the other under self-quarantine to prevent the spread of pandemic virus. Above all, the scarier thing must be demonic ideologies that separate us from a community of *salim* (enlivening).

At this juncture, what we need is indeed "healing." *We all need to become mothers of others* in terms of fecundity as we are all evolving together. Eradicating a certain virus might be impossible. We as the vulnerable community of *salim* need to cast out demonic delusions of greedy ownership over nature, which is the fundamental mentality of demonic ideologies: racism, sexism, heterosexism, speciesism, colonialism, imperialism, and Trumpisms. I am hoping for a community of interpreters, a great mother with Eros of life, a *salim* community. Going beyond a feminist essentialism, I truly believe that this time also is returning to be in a life cycle of yin-yang as an encompassing ultimate of which we all are mothers of the other as we all are mirroring each other in the process of *nature naturing* and *nature natured*. We, the (therapeutic) community of interpreters, recognize/conscientize pains and dangers of living together, take a side of love, and conjointly enter into the cycle of symbiosis with others for healing this *Life* community, which is called *Nature*.

NOTES

1. Robert S. Corrington, *Nature and Nothingness* (London: Lexington Books, 2017), 156.
2. Ibid, 150. All epigraphs in this chapter from Robert S. Corrington's, *Nature and Nothingness* (London: Lexington Books, 2017), are reprinted with permission. All rights reserved.
3. The epigraphs of *Dao De Jing* in this chapter are my translation. *Dao De Jing* 道德經 is a Chinese classic text traditionally credited to the sixth-century BCE sage Laozi 老子.
4. "Mission and Vision." Womensmarch.com. January 25, 2017.
5. Corrington, *Nature and Nothingness*, x.
6. Ibid., 125.
7. Hannah Arendt, *The Human Condition* (Chicago: University of Chicago Press, 1998), 199.
8. Ibid., 198.
9. Judith Butler, *Notes Toward a Performative Theory of Assembly* (Cambridge: Harvard University Press, 2015), 72.
10. Ibid., 156.
11. Corrington, *Nature and Nothingness*, 27.
12. *Dao De Jing* 道德經, Chapter 75.
13. Corrington, *Nature and Nothingness*, 28.
14. Ursula Goodenough, *The Sacred Depths of Nature* (Oxford: Oxford University Press, 1998), 11.
15. Ibid., 13.
16. Robert S. Corrington, "Evolution, Religion, and an Ecstatic Naturalism," *American Journal of Theology and Philosophy 31*, no. 2 (May 2010): 129.
17. Jea Sophia Oh, "Delivering Peace Out of the Broken Womb," *Current Dialogue 55*, Special Issue for the 10th Assembly of the World Council of Churches (September 2013): 29.
18. Ibid., 29.
19. Roger T. Ames and David L. Hall eds., *Dao De Jing, A Philosophical Translation* (New York: A Ballantine Book, 2003), 194–195.
20. Michel Foucault, *The History of Sexuality Vol. I: An Introduction*, trans. Robert Hurley (New York: Vintage Books, 1978), 94–95.
21. Gilles Deleuze, "On Capitalism and Desire," *Desert Islands and Other Texts*, trans. Michael Taormia (New York: Semiotext, 2004), 263.
22. Ibid., 267.
23. Corrington, *Nature and Nothingness*, 66.
24. *Dao De Jing* 道德經, Chapter 6.

25. Sara Ahmed, *The Cultural Politics of Emotion* (London: Routledge, 2015), 173–174.

26. Affects etymologically is from *adfectus* in *Latin*, as Spinoza discussed in his *Ethica*, states of mind and body related to feelings like pain and joy.

27. Gilles Deleuze and Félix Guattari, *A Thousand Plateaus: Capitalism and Schizophrenia* (Minneapolis: University of Minnesota Press, 2007), 256.

28. Benedict de Spinoza, *Ethica*, ed. and trans. Edwin Curley (Princeton: Princeton University Press, 1994), 113.

29. Deleuze and Guttari, 257.

30. Ames and Hall, 85.

31. Corrington, *Nature and Nothingness*, xi.

32. *The Candlelight Demonstrations* (2016–2017) were a series of protests against President Park Geun-hye that occurred throughout South Korea from November 2016 to March 2017. After the initial demonstrations on October 26, 2016, hundreds of thousands of South Korean protesters denounced the Park administration's political scandal and called for the resignation of Park.

33. Kim Ji-Ha, *Saengmyeong* [Life] (Seoul: Sol, 1992), 191–192.

34. Paulo Freire in his book, *Pedagogy of the Oppressed*, recognizes consciousness as something cultivatable and transformative. Conscientization means "critical consciousness for transforming the world," grounded in Marxist critical theory, that focuses on achieving an in-depth understanding of the world, allowing for the perception and exposure of social and political contradictions. When this happens in the process of learning, people discover that they are creators of culture for transforming the world. See Paulo Freire, *Pedagogy of the Oppressed*, trans. Myra Bergman Ramos (London: Continuum, 2005), 31. A feminist scholar and activist, bell hooks (Gloria Watkins) notes a link between "decolonization" and Freire's "conscientization" as critical consciousness. When it came to decolonization, education led to a "historical moment when one begins to think critically about the self and identity." See bell hooks, *Teaching Community: A Pedagogy of Hope* (New York: Routledge, 2003).

35. Corrington, *Nature and Nothingness*, xi.

36. *Dao De Jing* 道德經, Chapter 25.

37. Judith Butler, *Gender Trouble: Feminism and the Subversion of Identity* (London: Routledge, 1999), 19–20.

38. Jea Sophia Oh, *A Postcolonial Theology of Life: Planetarity East and West* (Upland: Sopher Press, 2011), 50.

39. Corrington, *Nature and Nothingness*, xi.

40. Robert S. Corrington, *Nature's Religion* (Lanham: Rowman & Littlefield Publishers, 1997), 23.

41. Robert S. Corrington, *Ecstatic Naturalism: Signs of the World* (Bloomington and Indianapolis: Indiana University Press, 1994), 9.

42. Corrington, *Nature and Nothingness*, 156.

43. Ibid., 19.

44. Ibid.

45. Ibid., 20.

46. Catherine Keller, *Facing Apocalypse: Climate, Democracy, and Other Last Chances* (New York: Orbis Books, 2021), 185.

REFERENCES

Ahmed, Sara. *The Cultural Politics of Emotion*. London: Routledge, 2015.

Ames, Roger, and David L. Hall eds., *Dao De Jing, A Philosophical Translation*. New York: A Ballantine Book, 2003.

Arendt, Hannah. *The Human Condition*. Chicago: University of Chicago Press, 1998.

Butler, Judith. *Gender Trouble: Feminism and the Subversion of Identity*. London: Routledge, 1999.

———. *Notes Toward a Performative Theory of Assembly*. Cambridge: Harvard University Press, 2015.

Corrington, Robert S. *Ecstatic Naturalism: Signs of the World*. Bloomington and Indianapolis: Indiana University Press, 1994.

———. "Evolution, Religion, and an Ecstatic Naturalism," *American Journal of Theology and Philosophy* 31, no. 2, 124–35, May 2010.

———. *Nature's Religion*. Lanham: Rowman & Littlefield Publishers, 1997.

———. *Nature and Nothingness*. London: Lexington Books, 2016.

de Spinoza, Benedict. *Ethica*, edited and translated by Edwin Curley. Princeton: Princeton University Press, 1994.

Deleuze, Gilles. *Desert Islands and Other Texts*, trans. Michael Taormia. New York: Semiotext, 2004.

Deleuze, Gilles and Félix Guattari, *A Thousand Plateaus: Capitalism and Schizophrenia*. Minneapolis: University of Minnesota Press, 2007.

Foucault, Michel. *The History of Sexuality Vol. I: An Introduction*, translated by Robert Hurley. New York: Vintage Books, 1978.

Freire, Paulo. *Pedagogy of the Oppressed*, translated by Myra Bergman Ramos. London: Continuum, 2005.

Goodenough, Ursula. *The Sacred Depths of Nature*. Oxford: Oxford University Press, 1998.

hooks, bell. *Teaching Community: A Pedagogy of Hope*. New York: Routledge, 2003.

Keller, Catherine. *Facing Apocalypse: Climate, Democracy, and Other Last Chances*. New York: Orbis Books, 2021.

Kim, Ji-Ha. *Saengmyeong* [Life], Seoul: Sol, 1992.

Oh, Jea Sophia. *A Postcolonial Theology of Life: Planetarity East and West*. Upland: Sopher Press, 2011.

———. "Delivering Peace Out of the Broken Womb," *Current Dialogue* 55, Special Issue for the 10th Assembly of the World Council of Churches (September 2013).

Part IV

NURTURING NATURE AND POSTHUMANISM

Chapter 9

Evil as Human Resistance to the Indifferent Force of the Primal Nature

On Storytelling Animals

Iljoon Park

INTRODUCTION

This chapter is about the signifier, evil, which is an empty signifier but points to something other than itself. The signified from the signifier "evil" is ontologically reticent but manifests itself sometimes as signs for pain and suffering and other times as unhappiness and dissatisfaction for sign users, especially in the narrative structures of human groups, in which the meanings of world, life and identity are explained. The linguistic sign, evil, is in this context human resistance to the indifferent force of the primal nature, which produces "holes in nature" and "totalizing nothingness" in Corrington's words. These are the linguistically expressed aspects of the primal reticent nature and reflect mainly human experience of pain and suffering. Ironically, the signifier, evil, points to something that lacks in human life but that drives the meaning of life home to human sign users. Thus, the signifier "evil" signifies both something that cannot be explained but experienced as bad or painful and other thing that is absent to actualize the fullness of life for sign users. The signified is delivered through stories in the narrative universe of the social imagery of people in a certain period and locality.

One of the problems of the traditional stories of evil is that they are based upon the binary structure of good and evil, friend vs. foe, in which the evil is always the bad side. This dualistic imagery of good and (or vs.) evil hides the real source of evil, although the dualistic structure is part of human effort to overcome the evil, which is as a matter of fact the indifferent nature. Instead of the simplistic dualism, human communities in the posthuman age need a new narrative structure by interpretive community, which generates the spirit of emancipation by producing the Two that points to the ungraspable dimension of the truth of nature and that gives birth to the virtual. By telling stories of evil, humans resist the indifferent and reticent nature in order to make human life meaningful in their narrative universe.

I will propose in this chapter that evil is construed as human resistance to the indifferent force of the primal nature which produces "holes in nature" and "totalizing nothingness."[1] Evil is produced by human interpretive resistance of storytelling in human life. In natural communities, evil is identified as the demonized other(s) and manifested with religious violence as "emotional plague" in Wilhelm Reich's terminology.[2] This ruthless violence in any natural community may be a response to individual experiences of "holes in nature," which are signified as "evil," although this demonization is not a healthy solution at all. As natural communities cannot find satisfactorily acceptable interpretations to address the experiences of radical indifference as "holes in nature," the meaning of nothingness is deeply cast down into a community, and such experiences may be understood as the "totalizing nothingness," especially when particular iterations of human beings "have a low tolerance for chaos and ambiguity."[3] In such instances the demonized other(s) become the sacrifice for the rage and despair felt in the face of totalizing nothingness as radical indifference to individual and social selving processes, and this sacrifice may in turn produce the fantastical delusions of safety and comfort. In this process, a natural community would turn into an aberrant interpretive community, that is, an evil one. Thus, this desperate experience has to be transformed with alternative interpretations unless the community in question would get trapped in a wicked circle of demonization, in which the community may ceaselessly try to find a demon to propitiate the dis-ease caused by certain disasters, accidents, misfortunes, or any experience that would be threatening to a selving process and meaning negating. In seeking to transfer the negative affects of guilt, shame, and blame onto a demonized Other, the fantastic logic of sacrificial propitiation is applied until the "evil cause" would seem to disappear.

The so-called posthuman age issues the question of evil with regard to human nature, especially when some belonging to *homo deus* would seek to enjoy eternal joy and happiness, while the others must live below the level of being-human, being treated as "superfluous" beings in a posthuman hedonic calculus. In such situations, how are we to interpret that the vivid manifestation of evil in our life has been changed from the issue of pain and suffering to that of unhappiness and dissatisfaction. In this interpretive shift, the human community of interpretation seeks for emancipation from the evil agents and structures by putting themselves in a state of *ec-stasis*, that is, by putting themselves outside the current vicious state into which they are stuck.

In natural communities, evil is identified as something substantial and agential, while, in active communities of interpreters, it is recognized as a mere signifier for something that does not exist in the established social order and that requires an interpretation. In a sense, any interpretation of evil can be a generation of fantasy since, "when fantasy and reality are one and the same, fantasy is not fantasy but reality."[4] As a signifier, evil ironically discloses that the sign is empty. However, this emptiness of the signifier or the sign is its real purpose, which finds resonance with the Daoist words found in the *Dao De Jing*, chapter 11, in which the purpose or use of a bowl does not lie in the bowl itself but in its empty space.[5] What it can contain is what it is absent for the present. According to Gregory Bateson, "the spoken word communicates something quite different from the body language."[6]

The uniqueness of human speech lies in its creation of "a distance which allows for an absence or, as it were, for a 'not.'"[7] This "not," a nonexistence or absence, lays the foundation of all signification. Thus, what one should ask is, "how can signification arise out of something that signifies nothing?"[8] Signification is none other than a reference to the absence or nonexistence. By referring to the "emptiness," it introduces a certain difference, which generates some ripples or lumps for signification. Just as it is "ripples or lumps" of energy from the big bang that produced difference and then existence of stars, planets and so on,[9] so semiotic beings are created out of semiotic nothing (or emptiness). In other words, the "not" or "emptiness" generates differentiation, which "mak[es] a hole in something that is continuous."[10] The absence of the object, to which the signifier refers, introduces a certain sense of lack in its signification and asks a change of the structure of signification. In this way, the nonexistence or emptiness of the signifier, evil, plays a role of human interpretive resistance against the continuous fabric of the existing structure, making a hole within it by producing a signifier for emptiness. Between the existing structure and the non-existing emptiness, the virtual is generated for signification.

EMPTY SIGNIFIER

When physicists found that elementary matter is constituted as a particle-wave duality, they were shocked and at a loss of words to adequately describe such quantum reality, for the mode of particle is totally different from that of wave and the equation of the former thus cannot be reconciled with that of the latter. Further, the duality is not like the phase transition of ice into water. The duality means that the particle state coexists at the same time as the wave state. Thus, a particle is at the same time a wave. It is as if the ice and the water modes emerge at the same time. It was at first thought as the measurement problem, as our linguistic and mathematical languages are abstractions from reality. However, the Copenhagen interpretation concluded that the duality is an inherent feature of reality and was later confirmed by Niels Bohr's complementarity principle and W. Heisenberg's uncertainty principle. The problem does not lie in the fact that reality is captured by two irreconcilable equations of particle and wave, but that each equation relatively works well with our cognition of reality, but each mode cannot confirm to the other mode. How is this possible? Then, what is reality after all? Particle or wave? Both? This is an impossible possibility, but the duality at the heart of quantum theory shows that this impossible possibility is actuality.

This changes our understanding of natural laws in that natural laws, mathematically expressed in equations, cannot reach the foundation of reality, but can only touch part of its superficial aspects. It seems then that Kant was right after all. Yet his epistemological categories were not exactly right. For example, his categories of space and time cannot stand against Einstein's concept of the time-space which says that time and space cannot be separated. Kant was right only in the sense that one cannot directly reach at reality. All our concepts, notions, equations and so on cannot escape from Whitehead's "fallacy of misplaced concreteness"[11] in that they are all abstractions from concrete experiences. That is, mathematical equations and abstract language for reality are empty signifiers, which are empty in themselves, but which respectively point to certain aspects of reality. One cannot say what reality is, but one can only partly apprehend it through the mathematical equations and languages, which show partly how it works. One can only guess the whole of reality by observing part of its aspects.

However, our understanding of reality in terms of mathematical laws is not as stable as we usually think of. For example, Newton's law of universal gravitation had to be re-interpreted by Einstein's theory of general relativity, but this reinterpretation was not an easy matter, for the former predicts that any change in the universe can instantaneously spread over the entire universe, while the latter predicts that the spread has a limitation of the light speed in that there is nothing faster than light. Thus, at a certain level, it requires a paradigm shift. What is a paradigm? It is like a narrative universe, in which one can reach understanding and successfully create and communicate one's ideas and thoughts with others. In this context, C.S. Peirce goes even further in saying that a natural law is "a cosmic habit that has emerged from earlier cosmic habits and thus exhibits more expansiveness and scope than its predecessor."[12] Whether one accepts the Peircean insight or not, natural laws or mathematical formulae are not some eternally existing set of objects that we have thought of. Rather, they are more like entangled stories which deliver us some of superficial aspects of reality. In other words, they are signs to be interpreted. They are the constituents for the stories of the universe or nature which we humans weave. Indeed, reality is told through stories that are the fabric of nature for human beings.

Evil is an empty signifier that tells us how nature or a world is experienced for humans. Although evil is a sign to signify something other than itself, it is common experience that the world, in which we live, is experienced as being full of evil. Some might say they are happy in this world, but it cannot be denied that there are so many innocent people suffering from evil. Yet, one cannot say that the world is evil or that nature is evil. Rather, it does not know any evil, because "evil" is a human concept, not a natural one.

Indeed, nature is indifferent to good or evil, but human beings live in a world always on the verge of suffering. For us humans, the indifference of nature is mostly experienced as our feeling of being meaningless in the world. It is as if one is abandoned without any care in the world. This would in turn generate the feeling of worthlessness or nihilism in our minds, and the latter in turn produces a sense of evil, usually in a way to point to the cause of the feeling of worthlessness in order for the person in question to deny one's meaninglessness.

Here I want to argue that the notion of evil is as a matter of fact simply human resistance to the indifference of nature by making an origin story for the indifference as a narrative of evil. It is human resistance to the indifference to find meaning in the world. Human beings are "the story-telling animals."[13] Humans tell stories to explain things in experience and to impose structures of meanings into and over natural events. In the stories, everything is interconnected, and all things have their own meanings in the narrative structure. Nothing is meaningless. Even pain and suffering can acquire meanings in people's life stories. The notion of "the evil" refers to the totality of the pain and suffering that any narrating self (selving process) directly and/or indirectly experiences. The evil is something "recognized" by humans, as a totality, which encompasses all their experiences of pain and suffering. It means that evil is a concept to explain the phenomena of pain in the suffering human subject.

However, it is not a philosophical concept to define its boundary and limitation. Rather, it is an ambiguous and vague conceptual feeling, saying that there is something evil to cause this pain to "me," that is, the suffering human subject. For example, for Hannah Arendt, evil was experienced as "more radical than expected," and as having to do with "making human beings as human beings superfluous."[14] Arendt found the cause of the evil in "the delusion of the omnipotence (not simply the lust for power) of an individual men," for "[if] an individual man qua man were omnipotent, then there is in fact no reason why men in the plural should exist at all—just as in monotheism it is only God's omnipotence that makes him One."[15] In Arendt's thought, the evil is used to explain her experience of something that "mak[es] human being as human beings superfluous." Given that this concept is available only for human subjects, one can ask why only human beings need this conceptual device.

Part of the answer to that question lies in the nature of human cognition. According to Lakoff and Johnson, human cognitive systems have an unconscious cognition system in which cause-effect corollary works with cognition.[16] When something happens, our cognitive system works in a way to think of its cause. However, because it is an unconscious system, it does not offer a linguistically clear explanation. An evil may be linguistically identified as the cause of the pain and suffering the subject is experiencing. This linguistic explanation of the evil consists of concepts metaphorically constructed.[17] As linguistic animals, humans need some explanatory narrative for what they think it is bad, and the evil is conceptually presented for it. In contrast, animal subjects also suffer from bad things, but they do not need any linguistic explanation, although they may have some bodily gestures to signify it. It does not mean that animals do not recognize evil things but that they respond to the bad happening without identifying it with evil. Only humans understand them as caused by evil agents or forces. In this sense, evil is a conceptual explanation of human suffering. In other words, evil is a human conceptual experience used to explain the (mental and/or physical) pain the human subject has suffered from.

Our dilemma with evil is that, although it is seen and experienced almost everywhere, "the intellectual resources available for coping with it" are very "sparse and inadequate."[18] Any attempt to deal with the problem of evil is mainly about "how to reconcile evil . . . with religious beliefs and convictions."[19] In this context, evil is usually represented as "everything that one hates and despises, what one takes to be vile and despicable, which is to be violently extirpated."[20] However, given our twentieth-century experiences of war, massacre, genocide, discrimination, racism, sexism and so on, this poverty of discourse on evil cannot be acceptable. Obviously, there is "something about evil that resists and defies any final comprehension."[21] This semiotically excessive feature of evil may disclose that evil is not a substance or agent at all but rather belongs to the realm of the virtual, which constitutes the foundation of the structural possibility of our thinking about the world as an ethico-religious intelligibility.

Markus Gabriel suggests that the world does not exist, for the world conceptually refers to something like a totality in which everything existing is contained, yet it is logically impossible for any concept or notion to contain everything in it.[22] Everything exists as they are, whether they are real or actual or possible or virtual. However, the totality of them is not conceptually possible. I suggest that the concept of evil presents a similar case. A sign can refer to it, but the signifier never entirely discloses the totality of the object to which it points. It only works within the order of the virtual. In other words, evil is conceptually required like an empty set in set theory, but it does not actually or physically exist out there. The concept of evil does not refer to anything substantial or concrete. It just refers to the totality of bad things and happenings. It does not just come out of the heavenly realm of ideas. Rather, it is just a linguistic device to refer to something beyond our conceptual grasp.

St. Augustine's definition of evil as the lack of good was a resistant idea against the Manichean dualistic understanding of good and evil.[23] This dualist understanding that took good and evil as separate and independent agents was not acceptable to St. Augustine because such an idea was an affront to the mainstream orthodoxy that resulted from the wedding of Roman imperialistic ideology with a triumphalist Christian theology. That is, it would end up in accepting the reality of the evil beyond the power of a supremely powerful and good God. For him, the notion of evil is just not real. Augustine's understanding of evil was meant to explain human sufferings and pains in their contemporary understanding of the world. For him, the evil was a conceptual and/or linguistic device to tell people that only God created and rules over the world, and the Evil as such does not exist. Here in the nonexistence of the evil, what St. Augustine was actually doing was not to construct a metaphysical concept or notion for the evil, but to tell a story about evil so that people could understand why bad things happen. That is, the notion of evil was deployed as a storytelling device to explain something evil beyond human understanding based upon its cause-effect schema.

THE POSTHUMAN EVIL?

As we are entering a new age called the posthuman age, which Yuval Noah Harari calls the age of "*Homo Deus*,"[24] the experience of evil seems to be changed from the experience of pain and suffering to that of unhappiness and dissatisfaction. Along with this change, one might need a different imagination and different narrative structures for evil. What has been really changed in this age of *homo deus* is that *homo sapiens* are changing humanity itself and also that they are trying to or believing to eliminate the figure of evil with scientific and technological innovations by overcoming the causes of pain and suffering, generating some feeling of unhappiness and dissatisfaction.

In this age of *homo deus*, the major problems of humanity like "famine, plague and war" seem to be nearing solution, although many people still are suffering from them. In 2012, 620,000 people died "due to human violence," whereas "800,000 committed suicide and 1.5 million died of diabetes."[25] In 2010, obesity-related illness killed "about 3 million people, whereas terrorists killed a total of 7,697 people across the globe."[26] These figures show that mental health and obesity are more dangerous than famine, plague and war (or terrorism) in this age. That is, the causes of our pain and suffering have been changed, and we thus need a new way of thinking about suffering. In other words, our conceptual structure of evil has to be reformed along with this sort of social and cultural changes.

In such circumstances, humanity now seeks for new goals such as "immortality, happiness and divinity."[27] In 2017, the movie *Ghost in the Shell*, which was originally a Japanese-made sci-fi cyborg-animation 공각기동대(*gok gakgidongdae* 空殼機動隊, Mobile Armored Riot Police in 1995), was on screen, and it shows how immortality might be actualized in a future world, a world in which the human mind is electronically extended over a digital network and the body is totally mechanized. The biological limitations of human mind and body would thus be overcome, and the mind (and even the body as cyborg) is hybridized with artificial intelligence through the networks. Yes, humanity is dreaming of controlling death through GNR technologies (genetics, nanotechnology and robotics). These days, we humans are already cyborgs because we are living with artificial prostheses such as eyeglasses, smartphones, SNS, artificial arms and the like.[28] The movie, *Ghost in the Shell*, shows a future, in which our body would be a prosthesis itself, our mind would be a downloadable, and the mind would be extended through digital networks. We know the philosopher Descartes, who Gilbert Ryle famously critiqued as making a category mistake of a "ghost in a machine" was a dualist believing that the reality consists of thought and extension, and his idea seems to be proven as true in the movie although in a subverted way, in which the mind is extended and totally interconnected.

Once death would be overcome, or once one just imagines it would be so, what would be at stake is whether one is happy. We may live here on earth, possibly forever, if we can replace our biological body with a mechanical one. Then, what really matters to all of us is how happily we're going to live. Any unhappy elongation of life would mean the same thing as a living hell. The easiest way to feel happy is to take pills for the feeling. In this posthuman age, one can take pills for the happiness feeling. For happiness from the contemporary scientific perspectives are "nothing but different balances of bodily sensations"[29] which can be controlled with medications or other biosynthetic means. In other words, by rewiring one's brain with medication or other technologies, one can produce the feeling of happiness, given that "our happiness is determined by our biochemical system."[30]

We, the posthumans, have a means to manipulate our biochemical system to produce the feeling we want. According to Harari, the number of children with ADHD has significantly increased, but it is not due to the increase of the symptom. Rather, it is due to the known fact that the medicines for ADHD have the effect of improving attention and focus. And these days "completely healthy kids take such medications to improve their performance and live up to the growing expectations of teachers and parents."[31] In this posthuman epoch, people seem to believe that "it would be more efficient to change the pupil's biochemistry"[32] than changing the education system. Such insights are not new, for it has long been known what alcohol, LSD and other similar drugs do for our brain chemistry. Governments and corporations know this well as they have used them for the manipulation of human souls, especially in this global network-based capitalism. Instead of trying to make the world better, people are now manipulating their brain chemistry and controlling their virtual reality, in which the narrative self of the brain functions. In this way, the brain world "mirror[s]" the social world.[33] In this mirroring of the biological and the social, Prozac "allows one to obtain these goods at a low cost, not only because this medication is not expensive, but also because it allows one to avoid the psychical cost of acquiring these values."[34] Here, our thought of evil seems to be guided by some power in that it is identified with our unhappy feeling, not with the structure, which causes the feeling. It seems that we are driven by a consumerist narrative of happiness, not by our sense of justice.

Further, humans are now upgrading themselves into superhumans like gods in ancient myths. Even human mind as well as human body are upgraded with artificial enhancement devices and equipment. It is not just a matter of upgrading external tools but this also involves enhancing the body and the mind in terms of "biological engineering, cyborg engineering and the engineering of non-organic beings."[35] In human's history, no technological revolution has ever tried to change humanity itself, but the posthuman innovation stands unique in intending to change humanity itself into the posthuman by merging with robots, computers, and the virtual worlds. It means that human beings can become mechanistic beings, possessing superpowers or super-abilities like gods. It may literally be the end of human being or *homo sapiens*. In this context, Harari says that homo sapiens are in a process of turning themselves into *Homo Deus*. This divinization process is not like the Christian God made flesh, but rather more like becoming a god as found in Greek and Hindu myths, in which gods have human-like minds with super-abilities "such as the ability to design and create living beings; to transform their own bodies; to control the environment and the weather; to read minds and to communicate at a distance; to travel at very high speeds; and of course to escape death and live indefinitely."[36] In so doing, *homo deus* now seeks for happiness and immortality through the mechanization of the body and the electronic digitalization of the brain. This image of god is quite different from gods in any previous age.

What then is evil in this posthuman age? Given that the technologically divinized beings seek for happiness and immortality, evil is anything that make people unhappy and mortal. It seems that anything we (post)humans cannot overcome is regarded as evil, which just means our fallibility and mortality. Now when it seems that almost everything is under control of our technological hands, we nevertheless feel something that inevitably escapes our controlling powers. For example, even when happiness is largely controllable in a posthuman age, most of us do not really feel all that happy. This feeling of pervasive dissatisfaction causes some anxiety for us. And this leads to thinking of what and why it is so. So, our selving stories begin. That is, the signifier, evil, seems to be again "used to silence thinking and to demonize what we refuse to understand."[37]

BEING-HUMAN AS STORYTELLING

However, the uniqueness of being-human does not lie in its intelligence or ability to use tools or consciousness or morality but rather in their "ability to connect many humans to one another."[38] In the past, evil had been told of as something that was the cause of pain and suffering but that was not overcome by human effort and ability. With technological revolutions and advancement, *homo deus* now seems to overcome the traditional causes of pain and suffering, while generating other configurations of evil. Thus, the evil never dies but changes itself into another configuration within an always evolving human-narrating universe.

Any feature of an individual human being cannot surpass an animal's mental and physical abilities including intelligence. The really unique feature of being-human is its capability to cooperate "flexibly in large numbers."[39] Human ability to cooperate on a large scale is "ultimately based on our belief in imagined orders,"[40] that is, the orders commanded by a divinity. The orders are "set of rules that, despite existing only in our imagination, we believe to be as real inviolable as gravity."[41] These orders are expressed and explained by (religious) stories. Thus,

> as long as all Sapiens living in a particular locality believe in the same stories, they all follow the same rules, making it easy to predict the behavior of strangers and to organize mass-cooperation networks.[42]

In this context, the secret of human cooperation lies in their ability to "invent and spread such stories."[43] However, story is not just the product of a fictional imagination. Rather, it is enmeshed with reality. Conventionally, one has constructed the order of reality based only on the distinction between objective and subjective realities. However, our modern binary understanding of subjective and objective realities has hidden the virtual dimensions of reality.

Virtual reality is not as real as objective realities; it is also not as vivid as subjective realities. It should not be confused with the concept of possibility that contains a sense of temporal sequence. The virtual is not something to be actualized in the future. Rather the virtual constitutes the structural foundation of our thinking, whereas the possible is something one expect to be actualized in the future.

Although the virtual lacks its objective realities, it exerts its actual power upon present beings by providing the structural possibility of thinking, as fantasy is not fantasy but reality when it is believed that fantasy is really real. The power of the virtual is clearly seen in the effect of stories. Subjective reality "depends on my personal beliefs and feelings," but the virtual is a way beyond the subjective. For example, God, money, nationalism, and the world are dependent upon our social belief in them, and they are not in our personal realm of mentality. The realm of the virtual is an "intersubjective level," on which entities "depend on communication among many humans rather than on the beliefs and feelings of individual humans."[44] As a matter of fact, our world runs mostly by these intersubjective entities including God and evil. These intersubjective realities like money, nation, state, individual and so on do not have any objective substance or value. However, once millions of people believe in their value, they become reality so that they can "use [money] to buy food, beverages and clothing."[45] Laws, gods, nations, market, and so on, all are constructed based on the virtual realities and run human's belief in them.

Story creates the virtual and makes the world run on the basis of it, for meaning is "created when many people weave together a common network of stories."[46] It means that meaning is invented through the common network of stories. Human beings exist in their meaning fields which are the people's network of stories. When many people think and act according to the stories, in which they commonly believe, the people can cooperate each other, sharing the same viewpoint. This is the worldview, through which people see the world and reality. Thus, stories are none other than "the web of meaning" which spins and unravels new meanings.

The products of the intersubjective web of meaning like laws, forces, entities, ideals, morality, corporate companies and so on exist "purely in [people's] common imagination."[47] Whereas an animal's life is confined to their objective world and their subjective experiences, humans "use language to create completely new realities."[48] It is none other than the invention of stories in which the new virtual is created. In this sense, humans cannot be reduced to hormones and neurons or to biochemical algorithms. Indeed, stories are the source of human ability to cooperate effectively in large scales.

Given that sapiens are storytelling animals, evil as a storytelling device does not mean that it is simply fictional and thus useless or ineffective.

Rather, the virtual realities being generated in stories are more effective to weave human groups into a more functional unit than any other means.

THE POWER OF STORYTELLING AGAINST BIOCHEMISTRY

The concept of evil does not refer to any substantial thing, but it explains the causality of happenings through concepts and stories. Evil is a human resistance to the indifferent flow of nature for gaining their meaning of life. Instead of blaming the indifferent nature, the concept of evil is invented to explain that life has a meaning for each one. For human beings, anything happening is thought to have its cause, but the indifference of nature cannot be thought to be the cause. We need an enemy to explain our pain and suffering. This explanation of evil is based upon "metaphysical comfort,"[49] which refers to the absolute foundation of our moral confidence, as Nietzsche said once. From this foundation, we see the actuality of pain and suffering. For example, following Carl Schmitt's politics, American politics has constructed its "enemy" to define its own identity. The constructed enemy has always been very ambiguous but precise enough to set up the boundary of being-American. In the process of constructing the enemy, it has manipulated and controlled the fear and anxiety of the people. Thus, the story of evil or enemy is very important to maintain human groups as a more effective and functional unit.

Ironically, humans invent the virtual and in turn are subjected to by the structure of the virtual:

> Human cooperative networks usually judge themselves by yardsticks of their own invention, . . . In particular, human networks built in the name of imaginary entities such as gods, nations and cooperations normally judge their success from the viewpoint of the imaginary entity.[50]

Without the virtual construction of the imaginary entities, human society would not work: "Without commonly accepted stories about things like money, states or corporations, no complex human society can function."[51] It is the narrative self that offers stories we need for life. The so-called Kantian turn, which refers to the impossibility to reach at the Thing, inversely points to the fact that one can reach at things and the world only in terms of human epistemological categorical structures. However, Kant does not know the reality of the virtual, which interweaves subjective experience and feeling from the objective realities into stories. We humans reach at the Thing in terms of the stories about the reality out of which evil comes.

Why do we need a story to explain our suffering in terms of evil? This question is like why humans need oxygen to breathe. This is who we are. Without stories to explain who we are, we sapiens cannot live, for it is stories that give us meanings. We do not know for sure whether human's conscious experience has increased their chance to survive as a species, but we exist as storytelling animals which have won the evolutionary survival game so far. Nonetheless, one thing is clear that story creates virtual worlds which are neither objective nor subjective and in which we humans see and understand the world through the virtual realities.

The most important thing about virtual reality is that it works beyond or against our biochemical algorithms, and this is why our love often works against a selfish gene's perspective. As a matter of fact, the concepts of God, world, justice, love and so on are used to explain things in our human world, and these explanations run against the stories of genetic algorithms in most cases. Thus, when one says that evil does not exist at all and that it is just a virtual reality to conceptually explain our pain and suffering, one cannot say it is useless if it is so, because the concept of the virtual is like a field of consistency on which humans gain the meanings they need for their lives. For example, the subjective side of being, that is, what one feels for oneself, is always somewhat resistant to its perishing. This subjective side cannot be studied by objective scientific studies, but it nevertheless resists Nature's eternal commandment of perishing. Our ability to tell stories can be a technique of resistance against our biological imperatives that are subject to perpetual perishing, and, in this way, only sapiens fulfill the work of the "subject-superject"[52] in Whitehead's term. In Corringtonian terms, stories and their virtuals are always and already ec-static in that they give us the structures of meaning and some ways to transcend them. As existentialism showed, existence is always and already ex-sistence, that is, out of the existing states.

EVIL IS NOT THE ANSWER, SO THE STORY BEGINS

By removing the meaning network of religious stories from our daily lives, humans regained the power they used to sacrifice for God, and they are entering into the new stage of *Homo Deus*. In modern times, human life is characterized as "a constant pursuit of power within a universe devoid of meaning."[53] The fuel of the modern pursuit of power is "the alliance between scientific progress and economic growth."[54] Here is the exemplary story about the evil in our posthuman age:

> When Ebola erupted in West Africa in the summer of 2014, what do you think happened to the shares of pharmaceutical companies that were busy developing

anti-Ebola drugs and vaccines? They skyrocketed. Tekmira shares rose by 50 per cent and BioCryst shares by 90 per cent. In the Middle Ages, the outbreak of a plague caused people to raise their eyes towards heaven, and pray to God to forgive them for their sins. Today, when people hear of some new deadly epidemic, they pick up the phone and call their broker. For the stock exchange, even an epidemic is a business opportunity.[55]

As Whitehead once said, "life is a robbery," and a robber needs "justification."[56] Nature is indifferent, and everyone has their own subjective stories which contain good and evil together. Their stories are a bit different from nature's story, if it has, for each individual runs for its own survival and reproduction. This race of survival and reproduction justifies their selfish behaviors. From this perspective, everyone is evil, and everyone is at the same time good if they cooperate with each other to survive. And evil is created to mark the boundary of the "us." The above story contains such factors.

The structure of modernity has made things worse, for it has been based on the myth of myth of ceaseless economic growth. The myth was a fact when the Western European imperialism rose high. However, when one prospects a post-capitalist age, the infinite growth of economy turns out to be one of the sources of modern evil. In this sense, communism was not an enemy to capitalism but rather a critical companion and friend in that it firmly believed that the real foundation of human life is "material" and that economic growth is essential for achieving a utopian society. To keep the economic growth working, more consumption has been encouraged without any ecological considerations for sustainability, and as a result all of us are witnesses to the age of climate change and climate chaos.

The secret weapon of modern capitalism and communism for the ever-growing economy has been science, which "helped discover fresh sources of energy, new kinds of raw material, better machinery and novel production methods."[57] Indeed, the overwhelming growth of capitalism was possible "by mobilizing concepts and tools that were initially developed largely autonomously in the theoretical sphere or in the domain of basic scientific research."[58] At least this alternative has worked until the current global ecological crisis. Thus, the modern binary division between religion and science does not reflect their actual relations but is the projection of the modern man's political story of evil to divide between the "us" and the enemy. However, science is still regarded as a "hi-tech Ark," like Noah's Ark, which has the potential to deliver us from any ecological disasters.

Is science then the evil for this posthuman age? Saying that evil is a narrative resistance against nature's indifference means that there is no absolute evil and that any evil and enemy are invented ones. Bad things happen all the time, but the concept of evil as a totality of disastrous happenings is invented to tell people why they were so. In this context, evil as a concept is to resist the reality of perishing. That is, evil is not the answer to the problem but rather a question to the hole of the existing structure of the world.

The Christian idea of incarnation does not focus upon the idea of resurrection, but on the death of God on the cross. As a matter of fact, it is a story of defeat. Our savior was defeated by the ultimate evil of death. However, this loss turns into a story of victory, not by triumphalist resurrection but by accepting his own death up to a point where the incarnate God was abandoned by God Himself. In the cry, "*Eloi, Eloi, lama sabachthani?*"—which means, "My God, my God, why have you forsaken me?," the God was divided and became self-alienated as two. Although perhaps no one actually saw his resurrected body, but the story of the resurrection was spread among his believers. And, as we know, the story won the game. This story of the defeat of the Son of God was not the end of the story. The story of resurrection was the virtual, in which the defeat turned into that of victory. This is the interpretation of the pain and suffering in the present, in which evil is introduced to tell us the meaning of the pain and suffering. The defeat is "God's suffering," but here the simple dichotomy of God and the evil does not work, for God, the Son of God, died for human sins and their sufferings. It means that "human history is not just a theater of shadows, but the place of a real struggle, the struggle in which the Absolute itself is involved and its fate is decided."[59] Here, the evil seems to be God as the Father who sent the Son up to the death on the cross. As a matter of fact, the God the Father is none other than God the Son. Indeed, "what dies on the cross is not only the terrestrial-finite representative of God, but God himself, the very transcendent God of beyond."[60] So, on the cross, God was divided into the Two. This is aporia for believers, because it seems that God "killed" Himself called His Son or "committed a suicide." The Christian believers cannot allow these possibilities at all. The event on the cross exemplifies the dilemma of evil, the true evil which cannot be solved at all by any human means.

CONCLUSION

Any identification of evil cannot be an answer to the problem it causes. It just points to the hole of the existing system of thought. Here, "with communities of interpretation," evil has the potential to begin turning into "more of a positive shipwreck than a negative one."[61] Interpretation is none other than ecstasy. By this I mean one can get out of the evil situation (or problem) with shared interpretations as inter-subjective selving narratives as healing cultures. Such narrative cultures provide hope for liberation from evil negativity. Ecstasy in ancient Greek means "to be or stand outside oneself" or the (current) state. In this sense, interpretation can bring ecstasy to community.

Thus, interpretation is none other than the spirit of emancipation that releases one from the shackles of evil. Spirit is "a virtual reality in the sense that its status is that of a subjective presupposition: it exists only insofar as subjects act as if it exists."[62] It is none other than a story of emancipation for the subject, which was captured by the evil structure and released by the emancipatory power of the Spirit of interpretation through the story. The posthuman age needs a story for itself, but many posthuman or transhuman stories are just perverted versions of capitalism which has devoured the story of democracy. Can we have another resurrection story or simply accept the end of human beings as homo sapiens?

NOTES

1. Robert S. Corrington, *Nature and Nothingness: An Essay in Ordinal Phenomenology* (Lanham, MD: Lextington Books, 2017), x.
2. Ibid., 1–2.
3. Ibid., 14.
4. Jesper Hoffmeyer, *Signs of Meaning in the Universe,* trans. Barbara J. Haveland (Bloomington: Indiana University Press, 1996), 8.
5. *Dao De Jing*, ch. 11 says, "We Throw clay to shape a pot, but the utility of the clay is a function of the nothingness inside it" (Laozi, *Dao De Jing, "Making This Life Significant": A Philosophical Translation* by Roger T. Ames and David L. Hall [New York: Ballantine Books, 2003], 91).
6. Hoffmeyer, *Signs of Meaning in the Universe*, 7.
7. Ibid.
8. Ibid., 3.
9. Ibid., 2.
10. Ibid., 10.
11. Alfred North Whitehead, *Science and the Modern World, Lowell Lectures, 1925* (New York: The Free Press, 1967), 51.
12. Corrington, *Nature and Nothingness*, 21.
13. Jonathan Gottschall, *The Storytelling Animal: How Stories Make Us Human*, Korean translation by Seung-Yung No (Seoul: Minumsa, 2014), 240.
14. Hannah Arendt / Karl Jaspers Correspondence, 165. recited from Richard J. Bernstein, *Radical Evil: A Philosophical Interrogation* (Cambridge, UK: Polity, 2002), 207.
15. Bernstein, *Radical Evil*, 207.
16. George Lakoff and Mark Johnson, *Philosophy in the Flesh: the Embodied Mind and Its Challenge to Western Thought* (New York: Basic Books, 1999), 171–196.

17. Lakoff and Johnson, *Philosophy in the Flesh*, 6.
18. Bernstein, *Radical Evil*, 1.
19. Ibid., 2.
20. Ibid., 3.
21. Ibid., 7.
22. Markus Gabriel, *Warum Es die Welt Nicht Gibt*, Korean translation by Hee-Sang Kim (Pajoo, SK: The Open Books Co., 2017), 20.
23. John Hick, *Religious-Philosophical Understanding of God, Humanity and Evil*, Korean translation by Jang-Saeng Kim of *Evil and the God of Love* (Pajoo, SK: The Open Books Co., 2017), 53, 60–61.
24. Yuval N. Harari, *Homo Deus: A Brief History of Tomorrow* (London: Harvil Secker, 2015).
25. Ibid., 15.
26. Ibid., 18.
27. Ibid., 21.
28. Andy Clark, *Natural-Born Cyborgs: Minds, Technologies, and the Future Intelligence* (Oxford: Oxford University Press, 2003), 4.
29. Harari, *Homo Deus*, 35.
30. Ibid., 39.
31. Ibid.
32. Ibid., 40.
33. Catherine Malabou, *What Should We Do with Our Brain?*, trans. Sebastian Rand (New York: Fordham University Press, 2008), 33.
34. Malabou, *What Should We Do with Our Brain?*, 51.
35. Harari, *Homo Deus*, 43.
36. Ibid., 47.
37. Bernstein, *Radical Evil*, x.
38. Harari, *Homo Deus*, 131.
39. Ibid., 131.
40. Ibid., 142.
41. Ibid., 143.
42. Ibid.
43. Ibid.
44. Ibid., 144.
45. Ibid.
46. Ibid., 146.
47. Ibid., 149.
48. Ibid., 150.
49. Richard Bernstein, *The Abuse of Evil: The Corruption of Politics and Religion Since 9/11*, Korean translation by Jee-Han Ryu and Hyun-A Jo (Seoul: Ulyuck Publishing House, 2016), 28.
50. Harari, *Homo Deus*, 176.
51. Ibid., 177.
52. Alfred North Whitehead, *Process and Reality: An Essay in Cosmology, Gifford Lectures Delivered in the University of Edinburgh During the Session*

1927-28, Corrected Edition by David Ray Griffin and Donald W. Sherburne (New York: The Free Press, 1978), 28.

53. Harari, *Homo Deus*, 201.
54. Ibid.
55. Ibid., 203.
56. Whitehead, *Process and Reality*, 105.
57. Harari, *Homo Deus*, 212.
58. Malabou, *What Should We Do with Our Brain?*, 41.
59. Slavoj Žižek and Boris Gunjević, *God in Pain: Inversions of Apocalypse* (New York: Seven Stories Press, 2012), 157.
60. Ibid., 171.
61. Corrington, *Nature and Nothingness*, 155.
62. Žižek and Gunjević, *God in Pain*, 171.

REFERENCES

Ames, Roger and David L. Hall. *Dao De Jing: A Philosophical Translation*. New York: Ballantine Books, 2003.

Bernstein, Richard J. *Radical Evil: A Philosophical Interrogation*. Cambridge, UK: Polity, 2002.

———. *The Abuse of Evil: The Corruption of Politics and Religion since 9/11*. Korean Translation by Jeehan Ryu and Hyun-A Jo. Seoul: Ulyuck Publishing Hose, 2016.

Clark, Andy. *Natural-Born Cyborgs: Minds, Technologies, and the Future Intelligence*. Oxford: Oxford University Press, 2003.

Corrington, Robert S. *Nature and Nothingness: An Essay in Ordinal Phenomenology*. Lanham, MD: Lexington Press, 2017.

Gabriel, Markus. *Warum Es die Welt Nicht Gibt*. Korean translation by Kim, Hee-Sang. Kyunggi, Pajoo: The Open Books Co., 2017.

Gottschall, Jonathan. *The Storytelling Animal: How Stories Make Us Human*. Korean translation by Seung-Yung No. Seoul: Minumsa, 2014.

Harari, Yuval Noah. *Homo Deus: A Brief History of Tomorrow*. London: Harvil Secker, 2015.

Hick, John. *Religious-Philosophical Understanding of God, Humanity and Evil*. Korean translation by Jang-Saeng Kim of *Evil and the God of Love*. Pajoo, SK: The Open Books Co., 2017.

Hoffmeyer, Jesper. *Signs of Meaning in the Universe*. trans. Barbara J. Haveland. Bloomington: Indiana University Press, 1996.

Lakoff, George and Mark Johnson. *Philosophy in the Flesh: the Embodied Mind and Its Challenge to Western Thought*. New York: Basic Books, 1999.

Malabou, Catherine. *What Should We Do with Our Brain?* trans. Sebastian Rand. New York: Fordham University Press, 2008.

Whitehead, Alfred North. *Process and Reality: An Essay in Cosmology, Gifford Lectures Delivered in the University of Edinburgh During the Session 1927–28.* Corrected Edition by David Ray Griffin and Donald W. Sherburne. New York: The Free Press, 1978.

———. *Science and the Modern World, Lowell Lectures, 1925.* New York: The Free Press, 1967.

Žižek, Slavoj and Boris Gunjević. *God in Pain: Inversions of Apocalypse.* New York: Seven Stories Press, 2012.

Chapter 10

The Posthuman and an *Advaya* Dialectic of Sacrifice

Ick-Sang Shin

INTRODUCTION

This chapter aims to reveal the meaning of the term "relationship" in the posthuman situation. There are two main ways of defining posthuman subjectivity.[1] One is an anthropocentric view of classical humanism, and the other is a post-anthropocentric view of anti-humanism. Despite the difference between the two, both positions will agree to understand the posthuman subject as a relational and interdependent nexus of experience. But a problem emerges precisely at this point: even if relationality appears with multiple names, such as interdependence, interconnection, belonging to a multitude, otherness, non-unitary subjectivity, nomadicity, becoming, and so on, is not the question about the specific mechanism or modality of relating remaining unexplained? In other words, is a relationship bound up with equal reciprocity, or not? If might it be a non-equal reciprocity, how could such reciprocity come about? Posthuman relationship inherits the character of relationship of life, as long as the posthuman through technological mediation is not completely separated from human elements or actively imitate human elements.

However, the relationship related to the maintenance of life is not interactive but one-sided. Life sustains and grows through the sacrifices of another beings. Even in the posthuman situations, the sublime of the forced sacrifice will continue. By examining the homeostasis mechanism of life, it can be found that the subject problem in the posthuman situation still touches the problem of forced sacrifice. Moreover, this mechanism is dialectical, and has an immanent transcendence feature in that it is thoroughly temporal. This chapter proposes that the non-equal reciprocity of the homeostasis mechanism of life is an *advaya* (不二的, *bul-ijeog*) dialectic of sacrifice. This dialectic will develop into a *messianic* sacrifice by being compared to

Agamben's dialectic of double negation. Through this, the concept of *ecstatic* self-transcendence is newly introduced.

THE BECOMING-HUMAN OF MACHINE

There are two main ways of defining posthuman subjectivity. One is the way of defining it as a self that is technologically mediated from the perspective of using technology to change the natural nature of human beings thereby overcoming their previous biological limitations—that is, being a means of enhancing the capabilities of human beings. The other way emphasizes relational and interdependent subjectivity as positively embracing the end of classical humanism triggered by recent developments in science and technology in an anti-humanistic subjectivist philosophical tradition. In the former view, an anthropocentric view of classical humanism has had a major influence on the definition of posthuman subjects. While in the latter view there is an effort to try and define the posthuman subject from a post-anthropocentric view of anti-humanism.

As Rosi Braidotti emphasizes, technological mediation is one of central factors to a new vision of posthuman subjectivity.[2] And perhaps all posthuman discourses, whether focusing on human enhancement or dealing with human subjects mainly through post-anthropocentrism, it is very likely they will be affirming the importance of technological mediation. What this implies is that human subjects will be conceived in a whole new situation due to the birth of posthuman subjectivity through technological mediation. If we regard posthumanism as an extension of human enhancement, we should interpret posthumanism from the viewpoint of expansion of human subject—that is, by considering posthuman subjectivity as an ecstatic extension of human *homeostatic* condition. But we will have to carry out this hermeneutic while recognizing that on the side of treating posthuman as a project of post-anthropocentrism any posthuman subject will be interpreted as a totally new subject that is not bound by the modern concept of the human. Therefore, while the former way resolves the relationship between machines and humans as a merely cultural problem, the latter way is inclined to address this dynamic as a nature-culture continuum.

Despite these differences, both positions will agree to understand the posthuman subject as a relational and interdependent nexus of experience. However, if the relationship is given normative primacy over the former idea of a foundational subject, the relationship appears to be a kind of "non-unitary subjectivity"[3]—that is, an ethical subjectivity that can take responsibility for itself in discourses of ethical accountability as a shared praxis of the subject and its community of interpretation.

But a problem emerges precisely at this point: even if relationality appears with multiple names, such as interdependence, interconnection, belonging to a multitude, otherness, non-unitary subjectivity, nomadicity, becoming, and so on, is not the question about the specific mechanism or modality of relating remaining unexplained? In other words, is a relationship bound up with equal reciprocity, or not? If might it be a non-equal reciprocity, how could such reciprocity come about?

This short chapter attempts to explain this single problem. What I have come up with for this purpose is the present situation of becoming-machine, the heart of technological mediation. Interestingly, it is the becoming-human of machine that is one of the driving bases of posthuman becoming-machine. One of the key reasons why some machines can transcend human beings is the fact that these machines are increasingly learning from humans and other organisms. In particular, artificial intelligence (AI) is preparing to connect to various nature-culture continuums with reference to the mechanism of the human brain. A posthuman nature-culture continuum thereby includes the nature-culture continuum of machines. As a simple example, Deep Neural Network (DNN) and Deep Learning represent the neural network of the human brain and introduce the ReLU (Rectified Linear Unit) function, a base function of signaling system of neurons, into its detailed signaling system. So why should we think silicon-digital technologies are incapable of instantiating the same networks that allow for the evolved carbon-based modes of consciousness found in human organisms.

Therefore, in order to reflect on posthuman subjects born by technological mediation, we still have to look at the human brain-body-environment continuum. Perhaps this is because actual technology is not catching up with the posthuman imagination of technological mediation. In any case, the becoming-human of machines is dominated by an approach to embodiment wherein the human mechanism is foregrounded and then applied to a particular goal (e.g., chess or Baduk). On the other hand, if human neurological mechanisms themselves are closely related to the *homeostatic* mechanisms from autonomic motor systems to emotion-feeling systems, then what is to limit a fully conscious AI from emerging? Moreover, from an evolutionary point of view, maintenance of the *homeostatic* mechanism is a primary goal of the neurological mechanism. Hence, the reflection on such deep evolutionary stability of the posthuman subject should be based on the human *homeostatic* mechanism above all. So, the following pages will be examining the concept of neurological *homeostasis* as a starting point for reflection on the concrete aspects of posthuman relationships.

HOMEOSTASIS: BETWEEN *HOMEO* AND *ALLO*

The first use of the term "homeostasis" can most probably be traced to French physiologist Claude Bernard (1813–1878). He first used this concept to indicate a certain *internal environment* of the body and its capabilities to pay attention to the stability of surrounding environments. From this realization he got an insight into the fact that "the internal environment is a necessary condition for a free life."[4] This concept was further reconceptualized in the 1930s by the American physiologist Walter B. Cannon (1871–1945), who used the word to "describe the mechanism that maintains the constancy of the bodily fluid component, body temperature, blood pressure, and other physiological variables within a narrow physiological range."[5] However, the changing external environment must be premised on the idea that the stability of the internal environment or the maintenance of physiological variables within the body stay within a certain range. In other words, the concept of *homeostasis* refers to the tendency to keep constant an internal environment in response to external changes—indicating a dynamic but persistent relationship between the outside and the inside of the body. In fact, the fields of biology and neuroscience focus on this point and offer operative definitions of *homeostasis*. For example, Eric R. Kandel uses Bernard's concept of internal environment to define and explain *homeostasis* as follows:

> The active maintenance of a relatively constant internal environment is called *homeostasis*. Constancy of the internal environment is the basis of the freedom of action we and other animals enjoy because it partially decouples our physiology from immediate external conditions and greatly extends the range of available habitats.[6]

According to this definition, the *homeostasis* mechanism is maintained by actively separating the inside of the body from its outside. In other words, the fact that the concepts mobilized to explain this mechanism are activity and freedom gives rise to the impression that *homeostasis* strongly separates the human body from an external environment. Let's take the case of thermoregulation and thirst relief as an example. The former is an example of negative feedback loops and the latter positive ones.

The human body maintains its body temperature within a narrow range of around 36.5 degree Celsius. To do this, the hypothalamus of the brain senses the temperature of its blood vessels, and also detects the external temperature that is transmitted through the skin. That is, this neurological biofeedback loop is always checking the outside and inside temperatures and actively responds accordingly. If the body temperature is lowered, the hypothalamus will raise the temperature by activating the autonomic motor system to speed up metabolism. In the opposite case, it will slow down metabolic activity and reduce heat production. The thermal sensory information of the skin might be directly transmitted to the conscious areas of the brain and get the body to perform a purposeful activity to maintain its temperature. Likewise, the hypothalamus might also carry a purposeful activity by transferring temperature information to other parts of the brain as well as the autonomic motor system. Some possible results of such activity are moving to a place where you can avoid cold or heat, wearing or taking off clothes, and so on. In this way, the body temperature control system maintains the *homeostasis* through a negative feedback acting in the direction of reducing the range of felt temperature changes according to the previous temperature change information.[7]

Similarly, feelings of thirst are motivated by the interaction of internal stimuli and external stimuli. When a person becomes dehydrated, a physiological error signal such as lethargy or loss of appetite appears, which is to say, a motivation state due to internal stimuli. External stimuli are caused by the purpose created by the motivational state. For example, a person who wanders through a desert in a thirsty state will prefer to find an oasis over all other activities. Even if the person is hungry, if the person meets an oasis, the oasis will stimulate the person's thirst more and the person will be led to resolve thirst first before finding food. Such stimuli are an external stimulus. These stimuli interact with each other to form a positive feedback loop that helps the thirst relief be fully attained by not reducing but strengthening the thirst till being relieved.

As you can see in the above examples, to maintain a relatively stable internal environment of the body, the ensemble of *homeostasis* mechanisms "responds sensitively to the changes in the surroundings and adjusts the variables of each time to balance without bias."[8] *Homeostasis*, then, is a mechanism that constantly connects to the outside to secure an internal environment separate from the outside. It is the property of the *homeostasis* mechanism that separating and connecting are strongly correlated with each other.

In fact, the nature of the *homeostasis* mechanism is precisely the nature of complexity system. Complexity system is a system in which many elements are formed by nonlinearly interacting. These systems openly interchange information (or energy) with the outside and dramatically form order at the edge of chaos. The *homeostasis* of life is a representative example of the manifestation of the characteristics of complexity system. It "is possible because life is constantly exchanging information from the surrounding environment."[9] Thus, even if we understand the *homeostasis* mechanism of life in terms of physical complexity systems, it results in the interaction between the inside and the outside of life.

However, the general fact that *homeostasis* represents an organism's ability to maintain its system stably through continuous interaction with the outside should be examined in a more detailed manner. This is because the phenomenon of life is nonlinear and complex. The concept of *homeostasis* emphasizes the holistic features of life. These features do not only play elegant concerto but also make a loud noise of the streets. The *homeostasis* system that has various subsystems may be subject to extreme circumstances that cannot be sufficiently manipulated by the mechanism of its own. The *homeostasis* mechanism does not hang on the negative feedback loop, which goes back and forth in a kind of bouncy way within a given range. It also uses a positive feedback loop that works insofar as it puts an appropriate pressure on the biological system, when it is out of a given range, or when it is expected to be out of that range. This process of ensuring viability in the face of challenges and changes demanded both inside and outside the biological system is something we can call *allostasis*.

While *homeostasis* refers to maintaining a constant stability state, *allostasis* refers to being placed in various stability states. Of course, *allostasis* exists for *homeostasis*. However, *allostasis* does not simply follow or mimic the *homoeostasis* syntax for achieving *homeostasis*. In other words, it does not mean maintaining a state but rather references the "achieving viability through change of state."[10] Peter Sterling and Joseph Eyer define *allostasis* as "the process by which an organism achieves internal viability through bodily change of state. Allostasis comprises both behavioral and physiological processes that maintain internal parameters within the limits essential for life."[11] Bruce McEwen argues that *allostasis* can be viewed as part of a broader range of fight-or-flight response and is a concept of emphasizing "the point that allostatic systems help keep the body stable by being themselves able to change."[12] We fight against a given environmental change or flee if we do not seem to have a chance to fight. Such behavior is a process of changing ourselves in accordance with the environmental changes, and also can be a process of keeping ourselves viable through this change.

But there are two more important things to keep in mind within this dynamic of internal-external bodily attunement to an always newly emergent bio-cultural ideal of *homeostasis* through *allostasis*. One is that although the *allostasis* mechanisms are related to both the brain and the whole body, the source of this mechanism is above all the brain. Physiological changes "are driven by the brain's attempt to maintain internal viability amid changing circumstances in diverse environments."[13] The other is the fact that this attempt, at least in the short term, means that the *homeostasis* mechanism has failed to cope with the given problem.

Based on the discussion so far, the concept of *homeostasis* can be summarized as follows:

1) *Homeostasis* is the interaction between the inside and the outside of the biological system, separation and connection, stability and change, negative feedback and positive feedback.

2) The interaction of these binary-oppositions acquires concrete meanings through the ongoing and shifting relationships between *homeostasis* and *allostasis*. *Homeostasis* overcomes and maintains itself through *allostasis*. This is always a failure-based actualization.
3) Thus, these binary-oppositions in the *homeostasis* mechanism can transcend themselves immanently amid themselves. In other words, they can ecstatically go beyond their limiting oppositions without any nihilations of the basic dipolar configuration.

Thus, this dynamic *homeostasis* mechanism of life-preserving vitality can be said to be a dialectical mechanism. And the non-dualistic relationship, that is, the actualization of qualitative stability through qualitative change, makes *homeostasis* readable as a dialectical development.

ADVAYA DIALECTIC: FROM ETERNITY TO TIME

So, is *homeostasis* then basically the same thing as a kind of Hegelian dialectic based on "being"? If it is a Hegelian dialectic, there should be no outside. Hegel's dialectical movement of thought represents an enterprise based on reason's real ability to reach itself through a synthesis of historically posited theses and antitheses. It is considered to be the self-development of the sui generis as pure "being," presupposing that this self-development itself is the very unfolding of truth. Therefore, "nothing," which is in eternal conflict with "being," should not be considered to be wholly outside of "being" but rather inside of it. The reason why this conflict must be intrinsic to being is that both "being" and "nothing" are all indeterminate as the objects of pure thought which transcends finite and linear spacetime.[14] Thus, the contradictory interdependence between the two opposing phases in Hegel's dialectic is implied in the normative definition of these two phases, but not as a real relationship. When Hegel speaks of becoming in the realm of logic beyond spacetime, this becoming is just a metaphor, and not an actually observable entity. Hegel's dialectic is not appropriate then for characterizing the dialectical nature of *homeostasis*, in that it is a universal dialectic that excludes the possibilities of any outside of an onto-thesis and does not directly deal with the concept of becoming as a temporal occasion. However, the various kinds of tumultuous *élan vital* that surge up here on the Earth are never those of closed systems but always of open, dynamic systems.

So, is *homeostasis* a kind of double negative or "non-dual" *advaya* (不二的, *bul-ijeog*) dialectic based on the radical emptiness of sunyata as "nothing" at all? This kind of dialectic can be referenced to the theory of the non-dual (不二論, *bul-iron*) that developed in the context of Nāgārjuna's Middle Way (中道, *jung-do*). Since the character of the *advaya* (non-dual, 不二的) expresses a kind of dialectic in itself, the review of it is vital to understand the double negative *advaya* dialectic. First, let's start by considering the adjective, *advaya* (不二的).[15]

The Sanskrit adjective *advaya*, which corresponds to "non-dual," reminds us immediately of *Advaita Vedanta*, one of many Indian philosophies, because the Sanskrit noun "advaita" means "non-duality." In simple terms, it is only the difference between adjective and noun. But this difference is surprisingly significant. The difference gave rise to a decisive way of thinking and distinguishing Buddhism from Hinduism during the time of Gautama Buddha, and later gave rise to Mahayana Buddhism, which introduces new metaphysical speculations that in some ways goes beyond previous Theravada Buddhist philosophies. Mahayana can be conceived as a philosophical movement of returning to the original spirit of Buddha in the time of Nāgārjuna who inherited this way of thinking.

The philosophical circumstances are as follows. If using the term *advaya* is a case in which it is thoughtfully expressed in the view of negating the self-nature of all beings in the world, using the term "advaita" is a case in which only the selflessness as non-duality is expressed in the view of affirming itself even though negating the self-nature of all other beings. Therefore, the former becomes the principle of self-negation, but the latter becomes the principle of self-affirmation. Thus, the former moves to the middle way of "neither identity nor difference (不一不異, *bul-ilbul-i*)," the latter converges to the monism of "unity of Brahman and self (凡兒一如, *bum-a-il-yeo*)."[16] What I am pursuing here is the very former, the *advaya* way of thinking, which makes self-negation based on selflessness a principle of thinking.

According to this way of thinking, every being-in-the-world is always generating and disappearing everywhere—it does not stay in a simply located point or area of spacetime, this is the idea of "non-abiding" (無住, *mu-ju*). Self-negation then is a kind of process always in motion. To explain this, let's assume both ends of a dichotomy, self-nature and selflessness. Here, self-negation must be a movement that transcends by traversing these ends. In dualist thinking, however, the result of continuous self-negation will be a repetition of choosing between self-nature and selflessness. On the contrary, in the case of *advaya* thinking, continuous self-negation is a movement and envelopment which embraces and transcends the alternation of self-nature and selflessness at the same time. This movement forms a four-dimensional spacetime continuum across a three-dimensional space. This is a situation that is caused by the fact that *advaya* thinking not only rejects the ontological principle of an excluded middle but also rejects the logical principle of double-negation.

The principle of excluded middle says that ∀*A*, the proposition "*A*∧~*A*" is impossible. If this is denied, then "∃*A*, *A*∧~*A*" must be possible. And in this case, some kind of third beings that cause fissures on the joint surfaces of "*A*" and "not *A*" should appear. But, if something is "*A*" and "not *A*" at the same time, it can be neither "*A*" nor "not *A*." These beings then can form a dis/continuous sequence with a diverse spectrum lying between "*A*" and "not *A*." The self-negation of *advaya* thinking then is realize through a process of capturing the diversity of this sequence. It is an on-and-off or back-and-forth movement between fixed self-affirmation and fixed self-negation. So, at first glance, this movement is easy to be misunderstood as one-dimensional motion. A simple on-and-off, binary movement that can be seen as a tedious repetition always in the same lines. But this is a deep misunderstanding that loses sight of the fact that these third beings are "qualitatively" different from "*A*" or "not *A*."

The principle of not overlooking this qualitative difference is derived from rejection of the principle of double-negation. The principle of double-negation says that "∀*A*, ~(~*A*)=*A*." And this principle is also based on a false dichotomy. If this principle is denied, "∃*A*, ~(~*A*)≠*A*" should be possible. But if there is a case in which "not-not *A*" is not necessarily made equivalent to "*A*," it can be said that "not *A*" must be an independent entity that does not depend on "*A*," and furthermore that the consequence of such double self-negation must come from some "process" that does not always simply return to the self as onto-logical sameness. And this gives reason to think that such a process is more than a monotonous on-and-off movement between the self-affirmation of "*A*" and the self-negation of "not *A*." We can refer to this process as a rotational movement that draws the circle's trajectory.

As shown in the pictures below, the denial of excluded middle can be compared to a linear reciprocating movement, which is the result of looking from the side of rotational movement (figure 10.1). The linear reciprocating movement seems to move between varieties that do not have qualitative differences because it is a movement along with the equivalent points on a straight line. However, the original rotational movement (figure 10.2), to which the denial of excluded middle and double-negation is compared, forms a curvature out of a straight line and expresses qualitatively different diversity at every occasion of the movement. Moreover, rotational movement is not only dramatic in that it expresses qualitative diversity; but, more generally, rotational movement has two directions, clockwise and counterclockwise, so a method that can distinguish between these two can come into view. In physics, we distinguish between the two by considering the fact that the plane created by the rotational movement has two sides. We determine the direction through the so-called right-hand rule. When you grasp the fingers of your right hand except for the thumb in the direction of rotation, the direction of the thumb

Figure 10.1 Linear Reciprocating Movement of the Denial of Excluded Middle (Viewed from the Side of Rotational Movement).

Figure 10.2 Rotational Movement of the Denial of Excluded Middle- and Double-Negation.

is the direction of rotation. This definition is also dramatic because rotational movement requires three-dimensional space to describe its movement, even though it is a two-dimensional plane movement. In principle, in the two-dimensional plane movement, the third-dimensional direction remains as an area of nothingness, a dimensionless dimension as it were. However, the two-dimensional rotational movement should always be described only in the three-dimensional space because it is such a movement that can be explained only when assuming the third-dimensional direction. The 2D plane of being is manifested through the 3D space of nonbeing. Thus, the two-dimensional rotational movement cannot find its foundation in a two-dimensional plane world. The movement always embodies itself in a domain of nothingness beyond itself.

But there is one thing that remains unresolved in the explanation of *advaya* thinking in analogy to rotational movement. Even though rotational movement is a movement across a diversity of qualitative differences, it still appears to be iterative. For example, is there any change in the rotational movement that goes around one wheel and returns to the origin? 360 degree and 0 degree are not different. But this is to overlook the fact that all movements are always temporal. All movements are temporal, and time is based on change, so all beings must be transient. This fact is well expressed in what Martin Heidegger (1889–1976) states, "Dasein always exists in the existential mode of its possible time. Dasein is time, time is temporal."[17] In this case, it can be said that he has a command of *advaya* thinking (of course, he would not even know the concept of *advaya*) in that he understands the possibility as a possibility that cannot be indicated by anything and the concept of which is disappeared. In temporal movements passing through transient and provisional things, 0 degree can no longer be equal to 360 degree. Such a movement on the four-dimensional spacetime is an advance to make all concepts transitory. So, when every rotational movement also appears in conjunction with time, the movement is an advancing movement, a spiral movement. And this fact explains the meaning of the rejection of the principle of double-negation.

The discussion so far, which unfolds *advaya* dialectic into a world of time, provides a proper way of explaining the dialectical nature of *homeostasis*. Including temporal evolution is itself a dialectic. The ideal of going out of time to make a qualitative leap is often ignoring the fact that dialectic itself is a thinking way based on temporal experience. Dialectics, however, is also a way of explaining the actual occasions within time, as well as a movement that allows the ecstatic self-transcendence of these occasions within time. As long as the *homeostasis* of life deals with the occasions of the life as a result of temporal development, this homeostasis follows a way of *advaya* dialectics.

HOMEOSTASIS AND AN *ADVAYA* DIALECTIC OF SACRIFICE

Let's return again to *homeostasis* and think about it afresh. *Homeostasis* embodies the self-identity of life through differences from the outside. It is only through the interplay of the internals and the externals of life, of separating and connecting, of stability and change, of negative feedback and positive feedback, that a certain categorical stability, that is, *homeostasis* is achieved. This stability, this self-identity, is, in fact, only a tentative occasion in the evolutionary process of a dynamically changing rather than an invariant *being*. *Homeostasis* is a kind of non-entityness (*niḥsvabhāva*) and therefore is a kind of emptiness (*Śūnyatā*). When this fact is emphasized, *homeostasis* comes to be based on the interdependence in the theory of Karma (*pratītya-samutpāda*), and therefore, it may be thought that, as long as *homeostasis* is a general characteristic of the life system, "interdependence is the essence of all ecological relationships."[18]

But this is an over-generalization. The interaction of *homeostasis* serves for one side. Life phenomena are more unilateral than mutual. In other words, the interplay of the internal and the external of life or of separating and connecting is wholly related to the way in which various environments contribute to life itself, but not to the way in which life itself contributes to the environment.

The insight of Erwin Schrödinger (1887–1961) that life "feeds on negative entropy"[19] reveals this fact. Negative entropy means a decrease in entropy, so life is a highly ordered system that is formed in a direction of decreasing entropy against the second law of thermodynamics. However, the second law of thermodynamics, the law of increasing entropy, is established in a closed system. It means that in a closed system exchanging any energy or information with its surroundings is completely impossible. Therefore, entropy, in an open system where exchanging energy or information with its surroundings is possible, may not be increased. Closed systems are systems that do not interact with the outside, while open systems interact with the outside, so entropy can be reduced without applying the second law of thermodynamics in life, a system that interacts with the outside. But it is possible only in a way that causes or accelerates the increase of the entropy of the surroundings interacting with life. The decrease in entropy of one life is linked to the increase in entropy of some others. This is the essence of the situation in which the separating and connecting between the inside and the outside of life or between life and its environment interact with each other.

In other words, the high order of life is at the price of disorder of other beings. The maintenance and growth of one life are at the cost of death and destruction of other beings. This is the meaning of the vital interaction. To translate this simply into interdependence is then an excessive interpretive leap. Life phenomena should not be easily generalized by co-prosperity or mutual symbiosis. Rather, all life phenomenon should be viewed as based on a logic of sacrifice.

Co-prosperity or symbiosis is an ideology that fits the principle of mutual exchanges that supports market fundamentalism, which is the basis of neoliberal politico-economic systems. How absurd is the idea that the equal exchangeability of relationships in the market will create an equitable and reciprocal society of growing prosperity and happiness? *Homeostasis*, the constant stability of life does not indicate a world of equal exchange of equality. Rather, it tells the world in which all the conscious life that maintains itself based on the noble sacrifices of others should live to reflect on itself in this sacrificial sensibility.

The detailed description of the world of life that *homeostasis* informs can be found between a sameness and otherness, a *homeo* and an *allo*. The inside of constant stability is a busy world with boisterous and bustling changes everywhere. To realize *homeostasis*, life systems learn from their failures, seeks new ways toward viability, and are constantly adapting themselves to the changing environments and conditions by overcoming without ignoring the binary-oppositions of internal and external, of separating and connecting, of negative feedback and positive feedback. Life systems then are ecstatically self-transcendent in that they secure their stability based on changing and become "excellent operations (*energumen par excellence*)"[20] through maintaining and at the same time transcending themselves.

It is faith that Giorgio Agamben explains exactly with the words "excellent operation (*energumen par excellence*)." His reasoning for this conceptual convergence between faith and excellent operation can be traced to his reading of *Ephesians* 3: 7 (*κατὰ τὴν ἐνέργειαν τῆς δυνάμεως αὐτοῦ*, BGT) and *Philippians* 3:21 (*κατὰ τὴν ἐνέργειαν τοῦ δύνασθαι αὐτὸν*, BGT). The term "dynamis" translated into the power of God is a potentiality, and it is active in the subject of faith as long as it is being continually actualized (*energeia*) by faith. Thus, faith is an actualization of potentiality, and this is a transformation of chronological time into messianic time.[21] Such faith and "heavenly citizenship" also develops in chronological time, as long as it is connected with a certain expectation of messianic salvation. However, messianic time follows "a logic . . . in which the A / non-A opposition admits a third term which then takes on the form of a double negation: the non-non-A."[22] And this signals the third time that does not belong to the binary-opposition of *kairos* and *chronos*. The opposition of time and eternity transcends itself in time so that it leaps to a certain time that is neither time nor eternity.

The relationship between *homeo* and *allo* then precisely exemplifies messianic time. In other words, this biocultural vitalist dynamic of realizing embodied singularity as a durable subject shows a dialectic of double negation that operates in chronological time. Time transcending is time intrinsic.

Likewise, *allo* changes and transforms the intrinsic to a temporality that is leading the self-transcendent nature of a dynamic *homeo*. Dialectical movement takes place between *homeo* and *allo*, creating a space for sacrifice. In other words, the interaction of *homeo* and *allo* cannot occur in the form of a continual recurrence of the same or mere representation but appears as a spiraling, circulatory movement that advances along the central axis of rotational movement, whether positively or negatively, no matter what valuations are being made. And in this movement, it becomes clear that there is no way for any life to pay back the life-given sacrifices of all other beings. Sacrifice could not be compensated for or exchanged but can only be transmitted toward the future in the actual occasions of creative changes and ameliorative transformations.

Reflect upon sacrifice—and the fact that there is nothing we can do for the myriad victims for us will become evident in the movement of a dialectic of double negation. Do not stop there though but act according to the dialectic of freedom. In this action, we will be able to fully appreciate the other beings sacrificed for us and to meditate and prepare for our life to become an instance of an infinite sacrifice ceaselessly unfolding for all beings.

How miraculous that life itself and processes of homeostasis as dynamic equilibrium between an inside and an outside resembles the messianic. The cross of the messianic is a symbol of ultimate sacrifice in that it leads to the enlivening of other beings out of its kenotic suffering. One sacrifice demands another sacrifice to the beneficiaries of it, and so spreads life infinitely toward the future. It is for this that the Apostle Paul teaches: "if so be that we suffer with him, that we may be also glorified together."[23]

CONCLUSION: FOREBODING INTIMATIONS IN A POSTHUMAN ERA OF DIALECTICAL MOVEMENTS

The deep-learning mechanisms of new developments in technologies of AI imitate the way the human brain works, but the brain's neural network is inexorably bound to throwing itself into maintaining the *homeostasis* of its body. The human phenomena of intelligence and self-consciousness serve the purpose of actualizing the life phenomenon of maintenance of *homeostasis* in chronological time in the form of an *advaya* dialectic of double negation. This purpose is a provisional principle that leads to the further evolutionary process of life including the growth of the potentials for human beings. *Homeostasis* as a relational nature of life shows that the "working goal" of life itself always already overlaps with the "working ways" of life, as the *sacrifice*. However, AI does not yet have enough built-in neurological mechanisms that closely cooperate with and maintain *homeostatic* mechanisms leading to such a complex sacrificial logic of emotion-feeling systems. Nevertheless, the posthuman era is still coming.

In this post-post-modernist era, this situation reminds us of a problem of recognition continuing from modern to post-modern times. Without fully representing either modernity or post-modernity—a commitment to a dialectical movement of "change and transition in which something is not abandoned or sublated, another is grasped or reached"[24] is a basic mark of all ideals of progressive development. Now we are going beyond the modern times of "human" and post-modern times of "anti-human," and entering into the posthuman dialectical movement era, in which humanism and anti-humanism can neither be abandoned nor sublated, yet another new creation can still be grasped and reached.

But this era is not free from the criticism that a "cyborg's anti-representational body is just nothing but a reorganization system of late capitalist new body."[25] More so, as long as there is no proper answer to the question, "what kind of reciprocity does any relationship signify?" The fact that all sublime sacrifices for life are always thrown toward the future without returning to the past or present challenges the free market capitalism based on deceptive notion, that is, mutually beneficial reciprocity. The *advaya* dialectic of double-negation based on these sacrifices can serve as "an ancient future" in the posthuman dialectical movement era, which is not free from the dangers of techno-science capitalism with the free market on its back, maximizing the abundance of the present by consuming the future. This is why we should be more eager to reflect on the profound message of a messianic sacrifice and be acting according to an *advaya* dialectic of double negation than ever before.

NOTES

1. This work was supported by the Ministry of Education of the Republic of Korea and the National Research Foundation of Korea (NRF-2017S1A5B5A02025336). This chapter is an extension of Ick-Sang Shin, "The Posthuman and a Dialectic of Sacrifice," *Madang: Journal of Contextual Theology* 27 (June 2017), with major revisions and modifications.

2. Rosi Braidotti. *The Posthuman* (Cambridge and Maiden: Polity Press, 2013), 90.

3. Ibid., 93.

4. Eric R. Kandel, et al. *Principles of Neural Science*, 5th Edition, Vol.2, trans. Bong-Gyun Kang, et al. (Seoul: Bum-Moon Education, 2014), 1046.

5. Ibid.

6. Ibid., 1083.

7. For more details, see Kandel, *Principles*, 1056–1064.

8. Jong-Wook Kim, "Ecosystem and *Dharma-dhātu* as Complexity Systems," *Chul Hak Sa Sang: A Journal of Philosophical Ideas* 44 (Aug 2011): 24.

9. Moo-Young Choi & Min-Soo Kim. "Life in the Viewpoint of Complexity System," in *Quantum·Information·Life*, Hoi-ik Jang, et al. (Paju: Hanul Academy, 2015), 423.

10. Jay Schulkin. *Rethinking Homeostasis: Allostatic Regulation in Physiology and pathophysiology* (Cambridge, MA, and London: A Bradford Book and The MIT Press, 2003), 17.

11. Ibid., 21.

12. Bruce McEwen & Elizabeth Norton Lasley. *The End of Stress as We Know It* (Washington, DC: Joseph Henry Press, 2002), 6.

13. Schulkin, *Rethinking Homeostasis*, 17.

14. For a more detailed discussion of Hegel's "dialectic of becoming," see Hong-Gill Jo. "Meeting between Hegel's Dialectic of Becoming and Paticcasamuppāda Thought in Buddhism," *Journal of the Daedong(Great Unity) Philosophical Association* 49 (December 2009): 194–205.

15. The following which explains the term *advaya* is an excerpt from Ick-Sang Shin. "Religious Liberation and Elucidation of Being: The Non-Dual," *Theological Studies* 63 (December 2013): 137–141.

16. For an explanation and critique of the monistic feature of *Advaita Vedanta*, see Stafford Betty. "*Dvaita, Advaita*, and *Viśiṣṭādvaita*: Contrasting Views of *Mokṣha*," *Asian Philosophy* 20.2 (July 2010): 216–217, 219–220.

17. Martin Heidegger. *Concept of Time*, trans. Dong-Eun Seo (Seoul: Numen, 2011), 44.

18. Kim, "Ecosystem and *Dharma-dhātu*," 29.

19. Erwin Schrödinger. *What is Life?: The Physical Aspect of the Living Cell*, trans. In-Seok Seo & Sang-Ik Hwang (Seoul: Hanul Publishing, 1992), 111.

20. Giorgio Agamben. *The Time that Remains: A Commentary on the Letter to the Romans*, trans. Patricia Dailey (Stanford: Stanford University Press, 2005), 90.

21. Ibid., 82.

22. Ibid., 51.

23. *Romans* 8: 17, NIV.

24. Robert Heiss. *What is Dialectic?*, trans. Mun-Soo Hwang (Seoul: Seo Mun Dang, 1996), 55.

25. Dong-Jin Seo. "Body without Organs or Organs without Body—Representation of Body and its Crisis," *Literature and Boundaries* 4.1 (Feb 2004): 143.

REFERENCES

Agamben, Giorgio. *The Time that Remains: A Commentary on the Letter to the Romans*. trans. Patricia Dailey. Stanford: Stanford University Press, 2005.

Betty, Stafford. "Dvaita, Advaita, and Viśiṣṭādvaita: Contrasting Views of Mokṣha." *Asian Philosophy* 20.2 (July 2010): 215–224.

Braidotti, Rosi. *The Posthuman*. Cambridge, and Maiden: Polity Press, 2013.

Choi, Moo-Young and Min-Soo Kim. "Life in the Viewpoint of Complexity System," in *Quantum·Information·Life*. Paju: Hanul Academy, 2015, 393–440.

Heidegger, Martin. *Concept of Time*. Trans. Dong-Eun Seo. Seoul: Numen, 2011.

Heiss, Robert. *What is Dialectic?* Trans. Mun-Soo Hwang. Seoul: Seo Mun Dang, 1996.

Jo, Hong-Gill. "Meeting between Hegel's Dialectic of Becoming and *Paticcasamuppāda* Thought in Buddhism." *Journal of the Daedong(Graet Unity) Philosophical Association* 49 (December 2009): 193–215.

Kandel, Eric R. et al. *Principles of Neural Science*, 5th Edition, Vol. 2. Trans. Bong-Gyun Kang, et al. Seoul: Bum-Moon Education, 2014.

Kim, Jong-Wook. "Ecosystem and *Dharma-dhātu* as Complexity Systems." *Chul Hak Sa Sang: A Journal of Philosophical Ideas* 44 (Aug 2011): 7–36.

McEwen, Bruce and Elizabeth Norton Lasley. *The End of Stress as We Know It*. Washington DC: Joseph Henry Press, 2002.

Schrödinger, Erwin. *What is Life?: The Physical Aspect of the Living Cell*. Trans. In-Seok Seo and Sang-Ik Hwang. Seoul: Hanul Publishing, 1992.

Schulkin, Jay. *Rethinking Homeostasis: Allostatic Regulation in Physiology and pathophysiology*. Cambridge, MA and London: A Bradford Book, and The MIT Press, 2003.

Seo, Dong-Jin. "Body without Organs or Organs without Body—Representation of Body and its Crisis." *Literature and Boundaries* 4.1 (Feb 2004): 132–145.

Chapter 11

Education for the Symbiosis of Humans and Machines in a Posthuman Age

Eunkyoung Lee

INTRODUCTION: ARRIVAL OF A POSTHUMAN AGE

As the so-called "posthuman" (or "transhuman") age continues to arrive, contemporary science and technology, which have sought to enhance human abilities and even further to go beyond human limitations, are dreaming of overcoming biological human being by technological development. Here, the prefix "post-" of course does not reference a biological extinction of human beings; rather, it refers to the fact that human being as it becomes increasingly based upon technological mechanisms is in the process of transforming into a new digitally networked being with the help from the areas of artificial intelligence (AI), digital networks and virtual reality, and various other forms of cybernetically augmented reality technologies. So far, humans have tried to overcome their limitations within a horizon of biological boundaries, whereas as posthumans the very horizon of biological form can be considered a limitation to be transcended. In such an age, educational curriculums and systems, which have been based upon our understanding of human beings as biological individuals, no longer pertain to younger generations. Furthermore, when the definitions and concepts, which have hitherto regulated the normative and ontological boundaries of being-human are no longer appropriate for the coming posthuman age, in which humans symbiotically live with machines and the mind is extended over the entire world, our educational culture and institutions as a whole have fallen into a state of crisis. Thus, a most urgent task for the coming age thus will be a comprehensive educational reformation.

LIVING WITH MACHINES

Quite a long time ago now human civilizations entered the machine age, and now homo sapiens have reached what we might call a "Second Machine Age," in which human mental and intellectual powers are augmented, refined, and enhanced by a plurality of digital technologies.[1] A recently popularized term—the Fourth Industrial Revolution—seems to show the contemporary actuality, in which there is a pervasiveness of smartphones (hyper-connected society), big data (algorithmic society), sharing and gig economies (zero marginal cost society), GNR revolutions (genetics, nanotechnology, robotics), IoE (Internet of Everything) after the IoT (Internet of Things), maker culture, and the so-called intelligent information society—all deeply based upon AI.[2]

This coming of a Second Machine Age signals the rapid and radical collapse of the historical boundaries of human being, especially with regards to the boundary between the human body and machines. It implicates that we humans as embodied beings will have to learn how to live in this new intelligent information society, cooperating and cohabitating with machines. It further means that culture, education, and social environment, which surround us humans, will be transformed along with the flow of such changes. This changed future form of posthuman being will be in an increasingly symbiotic relation with machines. However, to be posthumans does not simply mean to be de-humans. Rather, it would mean that we humans will have "a new style of life" which will require new ethical values and newly "extended communal awareness," and this of course will include a sense of "territorial or environmental interconnection." It in turn foretells a new ethical singularity and solidarity, which will be very different from that of the modern classical humanism.[3] Such new somatic stylizations of being (post)human issue a need for a symbiotic ethics between humans and machines, and between humans and nature. G. Lloyd calls this "a collaborative morality,"[4] the normative criteria of which must foreground an "emphasis upon non-profit communality, acceptance of relatedness and its viral contagious feature, efforts to carry out experiment for possible choices and to realize alternatives, core role of creativity and a new interlink between theory and practice."[5]

The reason why we humans should learn how to live with machines is that, in the coming posthuman age, what has been regarded as particularly unique human sectors will be passed onto the sectors of AI and automated digital programs due to so many "brilliant technologies." Thus, the posthuman will have totally different environments of life and thus offer a different horizon for collectively interpreting the subjectivity and agency of persons beyond the classical inheritance of modern humanism as *homo sapiens*.

These days it seems many children hardly ever put their smartphones down. Previously we humans used to memorize schedules or telephone numbers in brains, such practices now seem to be a quaint memory of the past. Now persons utilize smartphones as their external memory devices.[6] Andy Clark calls this phenomenon "the extended mind," which means that, as humans carry out some cognitive tasks with external devices, our cognitive process is extended over the external devices, which in turn become part of our minds and/or cognitions.[7] In other words, cellular phone are no longer a simple communication device but rather becomes part of the human body, especially for children. David Chalmers considers how "iphone is already part of my mind."[8] Franco "Bifo" Berardi, who is an Italian Marxist theorist and a media activist, also emphasizes that machines, which can no longer be easily separable from our body and mind, both conceptually and existentially, become "no more simple external devices than internal transformation devices, which irretrievably modify human body and mind. In other words, the augmenting devices really amplify our bodily abilities and linguistic cognitive capacities."[9] In this sense then machines are nothing other than part of "ourselves."

Some years ago, the U.S. Whitehouse and the National Science and Technology Council announced seven principles for the imminent future in which human persons will live increasingly intimate with AI. The report, *Preparing for the Future of Artificial Intelligence*, suggests seven directions for policy-making for the time to come: (1) AI should seek to optimize public interests such as health care and public transportation and realizing a zero-crime environment; (2) the government should support AI research; (3) self-driving automobiles and drones should be actively regulated by lawmaking; (4) AI should not be conceived of as the replacement of humans but as a beneficial supplement to them; (5) one should not use any incomplete data; (6) no child should be left behind; and (7) always think first of security in the world. The general gist of the report can be summed up as saying that the use of AI should be regulated under public interests. And to achieve this purpose there is at the very least required a minimum amount of active legal intervention.[10]

Human symbiosis with machines offers a chance to recreate and restructure society and values through technological advancement. New technologies will "alter, enhance and, at the same time, extend our sense of presence and our potentials for creative action."[11] There is a sense that human civilization will fall into a huge crisis, which derives not only from the advancement of hi-tech technologies but also from the changes in human perspectives on technology and the new ways of conceiving relationships between humans and machines. In order to manage the coming crisis, we will have to leave behind our conventional attitudes and perspectives, which have predominantly regarded technology just as artificial prosthesis to supplement human frailty. We are asked these days to recognize technology as an experiential field that can actualize potentials to radically reform human societies and to leap into a new stage of being in history.[12]

FROM INDIVIDUAL SURVIVAL TO SOCIAL STRUCTING

A posthuman future cannot solve the problem of inequality. Given the high cost of human enhancement technologies, people who will gain some benefit from it would be the richest and the most powerful in developed countries, that is, the most prosperous 1 percent of capitalist expropriators. However, this should not entail an unconditional opposition to human enhancement technologies. For the issue of economic inequality is not only applicable to a posthuman issue; rather, many other problems in contemporary society have also contributed to amplifying the gap.

Michael Sandel suggests an ethics of giftedness, in contrast to technologies of human enhancement. In *The Case against Perfection*, he points out that human enhancement through biological engineering is nothing but the makeup for excessive anxieties for social-economic conquest and political dominion. Sandel argues that the anxiety to achieve ontological perfection with posthuman enhancements merely block and damage the potentialities of present life as overflowing with gracious gifts.[13]

In the same vein, Shin stresses that what is crucial is not human enhancement itself, but rather our consumer-driven capitalistic ways of life which recognizes only those with power and abilities and relatedly those who can purchase such powers and abilities itself.[14] One has to think about how the human enhancement technologies can influence our societies, institutions, environments, and whether technological enhancements will improve the quality of human life in general. These seem to be political, ethical questions, but we will have to ask these questions along with questions concerning technological development.

Entering into this posthuman age, we are overcoming many obstacles to human civilization, but we are at the same time losing so many things, which have made us happy. What we lost is human value such as kindness to strangers and caring for others. In this market-driven society, we don't have time for that. We are too busy. No time for coffee, I don't even have enough time to take care of myself. Nevertheless, we cannot live alone, even if human individual abilities and powers could be enhanced and upgraded with digitalized mechanical powers—for humans, after all, are social animals. The key to the human evolution has been an ability to weave people into some more cohesively and expansively set of united activities. With the emergence of larger-scale societies of cooperation humans can then become subjectivities actualizing their creative potentials in concert by overcoming their egoistic and atomized individuality by systematically constructing various forms of collaboration. For this purpose, nurturing a will to sacrifice oneself for the greater good of a larger whole is a critical development in the evolution of human morality. This is why altruistic behaviors are always encouraged in human societies. The role of religion has also certainly made a huge impact in this regard by encouraging an embrace of others as part of a divine-cosmic family and to give up themselves for an ultimate concern.

Peter Nowak in *Human 3.0* emphasizes a need for communal thinking or "collective thinking" to have any hopes of surviving in the posthuman society. This is akin to what Marina Gorbis calls "social structing" or the construction of social capital. Koreans call this human relationship (*ingangwankye* 人間關係), which includes kinship relations, personal relations, school relations, and so on indefinitely expanding into a cosmopolitan order. Human relations can become social capital. The point is not to manipulate human relations for selfish purposes but to make it function more harmoniously as shared social capital. Thus, the task of education in such a posthuman age is to empower students to construct their potential for generating social capital through nurturing of their healthy relations with humans and nonhumans.

According to Gorbis, contemporary individuals are amplified through communal intelligence and new technologies like social network services. "This amplified individual can change the world," says Gorbis. The social relations would become not only social capital but also the essence of the future world. The uniqueness of human being is their ability to connect with the many rather than their individualized intelligence or ability to make tools.[15] Sapiens have had ability to interweave people beyond the boundary of kinships, and this ability to connect has made sapiens prosper on this planet.[16] From an evolutionary perspective, sapiens do not merely excel other animals in intelligence, toolmaking, or physical capacities; rather, homo sapiens have been able to survive because they are the only species to gather many as a group and to cooperate flexibly with their in-group members.

In this train of thought, the notion of "the community of interpreters" in Robert Corrington's ecstatic naturalism offers a very fecund insight.[17] Having inherited a way of thinking about communal practices of understanding and meaning generation from the semiotic concepts of Charles S. Peirce, a founder of American pragmatism or "pragmaticism" in his own term, Corrington's notion of a community of interpreters points to possibilities for human beings as storytelling animals to fill a spiritual gap in the so-called "posthuman" age and to continue to seek out meaningful alternatives to nihilism and spiritual death. Stories and discourses that storytelling animals (ecstatically natural *homo narrator*) produce need some interpretative frameworks to be foregrounded productively to better weave them into a coherent guiding narrative for a posthuman future to be possible. According to Harari, religions have played such a role in human history. However, in the posthuman age, the traditional religious narrative drama does not seem to strongly attract attention anymore. Thus we need an alternative to weave a global narrative based upon personal and collective stories of cosmopolitan solidarity.

INCONVENIENCE AS POWER: CHANGING NOUN INTO VERB

Brilliant technological advancement has offered us innumerable benefits to overcome inconveniences and discomforts of all sorts, making our lives more at ease. This is the culmination of a history of human civilization that has sought for happiness by eliminating inconveniences and discomforts whenever possible. As a result, human life has been getting easier and more convenient in workplaces as well as homes. Our homes are full of various hi-tech devices, and we have gained unimaginable benefits from them. However, as we have been getting used to this comfortable life, some of us, especially people from currently in their tens to thirties, suffer from a mental symptom called digital dementia,[18] which can cause panic disorders, anxiety, and various forms of psychic and emotional disturbances. On the other hand, one might notice with increasingly alarm the pervasiveness of "discomfort marketing," which advertises discomfort instead of comfort for users or consumers. Such discomfort is marketed as needing human laborious participation instead of the totalizing full convenient services for users and consumers. This seems to be a basic inversion or perversion of the aforementioned notion of human history as perpetual progress into convenience and comfort.

Of course, convenience and comfort are not the devil themselves to be blamed for all of our social problems. To seek for them is a basic human instinct. However, it may become a totally different story that, as comfortable life easily turns into a main cause for modern diseases such as stress and depression, discomfort and inconvenience ironically awaken our survival instinct. In *Your Survival Instinct Is Killing: Retrain Your Brain to Conquer Fear and Build Resilience*, Marc Schoen and Kristin Loberg insist that comfort is a "wolf in sheep's clothing."[19] The comfort-oriented contemporary culture shows us a fast and easy way to satisfy our desires and derives, but this quick and easy satisfaction propels some disequilibrium in our mind. In this context, to utilize discomfort and inconvenience is our source of power to improve survival. Ironically, living with dis-ease becomes an advantage in the convenient culture.

As state-of-art scientific technologies brilliantly develop, our feelings of fear and helplessness proportionately increase. For our control over machines, which develops every day, becomes more and more weakened. Even the so-called digital natives and digital immigrants[20] feel that it is beyond their capacities to catch up to the rapid pace of technological developments. Kwang-Seok Lee calls the current overflowing hi-tech digital machines "the black box, which lies in the darkness far away from our controls."[21] In this context, one needs a shift of thinking to develop appropriate technology, which allows technology as only minimally as needed for our life, instead of following unconsciously, uncritically, and blindly the excessive flow of the advanced hi-tech developments. This is the tradition of ad hocism,[22] which sheds light upon a new wisdom of life by combining the past experience of technology with new ideas or inventions. Instead of making and purchasing new products, it recommends living with old ones by reforming them fit to changing environment. This points to the needs for more "fluent exchange and supplement between the old and the new beyond the inertia of individual technological activities."[23]

These kinds of ideas are already widely circulated in a field of educational theory called work education (*Arbeitserziehung*), which is a movement to recover human nature as alienated from work. As part of the Maker movement, this is now known as an alternative path to hi-tech civilization. The basic spirit of the maker culture is to rebuild and reuse discarded machines in a way as to optimize fit with alternative purposes. This maker activity corresponds with a mindset of "reverse engineering," which refers to activities engaged in to "understand closed commercial technology and design methods of machines and to newly re-design them for subjective purposes."[24] This self-maker culture to make, fix, and reuse things by oneself has some radical implications, especially in the regime of a globalized neoliberal economy, in which ecologically destructive and unsustainable consumerism is rampant.

The recently popularized "analog trend" also discloses the fact that the most advanced technologies and technological innovations are not always signaling progress. Analog experience can offer "the enjoyment and satisfaction the digital experience [armed with hi-tech devices] cannot give" and produce "some better achievement than the digital one."[25] Ironically, the innate weakness of analog technology can rather improve the productivity of users. Industry, especially automobile manufacturers, introduces a method of reskilling to overcome defects of automatically deployed workplaces. For example, Toyota, which once led the automatic manufacturing process, has again replaced some automatic robots on the assembly lines with skillful human labor worker, and this has led to the quality improvement of manufacturing.[26]

Ivan Illich criticized modern education as promoting infinite competition and individual survival to the detriment of more sustainable values of social and ecological solidarity. For him, part of the problem lies in our tendency to replace verbs with nouns. According to Illich, "Humans used to use verbs to describe their satisfactory actions, but they now replace them with nouns that have been invented for passive consumption."[27] His insight still has educational implications for our age of infinite competition and individual survival. In so much of contemporary education, words like learn, play, and rest get replaced with grade, game, healing, and so on. There is an impassable abyss between them. Getting grades, playing computer games, and seeking healing are things that an individual can largely do alone to a greater or lesser degree. However, the verbal activities of learning, playing, resting, one typically depends upon other bodies. As Diane Levin argues, "Bodily experience lies at the foundation of our relationship building,"[28] and this is an irreducibly basic element in all authentic learning as social, emotional, and cognitive developments. Thus, any effective education for the posthuman age should start with reclaiming bodily agency by changing nouns into verbs. It means that what we should resist is not against machine but against enclosed and static communities of interpretation and dogmatic cultures that nostalgically cling to the past.[29] Also, we as teachers should never confuse our satisfactory enjoyment of hi-tech devices and convenient digital machines with their educational effects.[30] Thus, we should never cease asking ourselves the following questions: "What should we teach children with these hi-tech devices?" and "Are we merely letting them enjoy the contemporary 'convenience and comfort' of technological advancements without developing any critical consciousness?"[31]

CONCLUSION: LIVING TOGETHER SYMBIOTICALLY

In this chapter I dealt with the educational aspects for the posthumans, considering the notion of social structing in terms of building up a social network between humans, appropriate technologies, and work education (*Arbeitserziehung*) and thinking of a possibility of education for the symbiosis of humans and machines in a posthuman age. Through this research, ironically, we found a paradoxical lesson that the more convenient and comfortable our lives will be, the more dependent we humans will be upon machines and their convenience. With this "the more . . ., the more . . ." logic there will be more chance for our minds to get manipulated—we might consider how the plot of the movie *Matrix* is seemingly becoming less incredible science fiction and more critical social commentary.

To live as a creative subjectivity in the posthuman age is nothing more than to realize this seductive trap of totalizing comfort and convenient technological services and to struggle "in a strenuous mood" with the inevitable discomforts and inconveniences in order to bring about a more significant and sustainable global community of interpretation. Living together with others symbiotically and interdependently, we can be "cultivating our morality" and bodily capabilities in order to use them more wisely.[32] In this sense, we cannot stop conversing about the intersections of technology and morality.[33] It is our imperative to be living for and with others, and wholeheartedly embracing discomforts and inconveniences.

NOTES

1. E. Brynjolfsson and A. McAfee, *The Second Machine Age: Work, Progress, and Prosperity in a Time of Brilliant Technologies*, Korean Translation by Han-Eum Lee (Chung-lim Press, 2014).
2. Kwang-Seok Lee, *Criticizing a Data Society* (Seoul: Chag-Il-Neun-Soo-Yo-Il, 2017), 61.
3. M. Nussbaum, *Frontiers of Justice: Disability, Nationality, Species Membership* (Cambridge, MA: Harvard University Press, 2006); recited from R. Braidotti, *The Posthuman* (Oxford: Polity, 2013), 190.
4. G. Lloyd, *Spinoza and the Ethics* (London & New York: Routledge, 1996), 74.
5. Recited from R. Braidotti, *The Posthumans*, 190.
6. Finlanders call cell phone *kanny*, and Germans call it *handy*. The word *kanny* is a childish word to refer to palm, and the word *handy* derives from the English word "hand" and means useful, convenient, and easy to handle. These expressions already imply that cell phone is not just a tool but rather part of our body.
7. Sang-Gyu Shin, "Extended mind and the extension of the measure," in Ewha Research Institute for Humanities & Labex Arts-H2H Laboratory, *The Theater of the Posthuman* (Paju, SK: Acanet, 2015), 36.
8. Andy Clark, *Supersizing the Mind: Embodiment, Action and Cognitive Extension* (Oxford: Oxford University Press, 2011), ix.
9. Franco Beradi Bifo, *After the Future* (Edinburgh, UK: AK Press, 2011), 22.
10. Executive Office of the President & National Science and Technology Council Committee on Technology, *Preparing for the Future of Artificial Intelligence: A Government Report* (Createspace Independent Publishing Platform, 2017).
11. Andy Clark, *Natural-Born Cyborgs: Minds, Technologies and the Future of Human Intelligence* (Oxford, UK: Oxford University Press, 2004), 114.
12. Jae-Hee Kim, *Techno-Philosophy of Gilbert Simondon: A Blueprint for the Posthuman Society*, Korean Publication (Paju, SK: Acanet, 2017), 12–16.
13. Michael J. Sandel, *The Case against Perfection: Ethics in an Age of Genetic Engineering*, Korean Translation by Su-Kyung Lee (Seoul: Wiseberry, 2016), 111–115.
14. Sang-Gyu Shin, *The Future of Homo Sapiens: Transhuman and Transhumanism* (Pajoo: Acanet, 2014), 241.
15. Yuval N. Harari, *Homo Deus: A Brief History of Tomorrow* (London: Harvill Secker, 2015), 131.
16. Ibid., 131.

17. Cf. Robert S. Corrington, *The Community of Interpreters: On the Hermeneutics of Nature and the Bible in the American Philosophical Tradition* (Mercer University Press, 1996).

18. Digital dementia refers to a symptom in which memory and/or numeracy is deteriorated due to our over-dependency upon digital devices like smartphone, PDA, computers, and so on.

19. Marc Schoen and Kristin Loberg, *Your Survival Instinct Is Killing: Retrain Your Brain to Conquer Fear and Build Resilience* (New York: Plume Books, 2014).

20. The term "digital natives" refers to younger generations raising up in digital-friendly environment, while the term "digital immigrants" refers to older analog generations being forced to get used to current culturally digital environment.

21. Lee, *Criticizing a Data Society*, 178.

22. Ad hocism is "the art of living and doing things ad hoc—tackling problems at once, using materials at hand rather than waiting for the perfect moment or 'proper' approach." Charles Jencks, *Adhocism: the Case for Improvisation* (1972), the front cover. This involves "using an available system in a new way to solve a problem quickly and efficiently." Charles Jencks and Nathan Silver, *Adhocism: The Case for Improvisation*, expanded and updated edition (Cambridge, MA: the MIT Press, 2013), vii.

23. Lee, *Criticizing a Data Society*, 137.

24. Ibid., 183.

25. David Sax, *The Revenge of Analog: Real Thing and Why They Matter* (New York: Ingram Publisher Service US, 2016), xvi.

26. Ibid., 158–159.

27. Ivan Illich, *The Right to Useful Unemployment and Its Professional Enemies* (London: Marion Boyars Publishers LTD, 1978), 6.

28. Sax, *The Revenge of Analog*, 180.

29. Kim, *Techno-Philosophy of Gilbert Simondon*, 19.

30. Sax, *The Revenge of Analog*, 182–183.

31. Ibid., 328–329.

32. Shin, *The Future of Homo Sapiens*, 140.

33. A good example of a topic that such a community of interpreters could take up is the urgent issue of Data Rights and/or the right to be forgotten.

REFERENCES

Berardi, Franco Bifo. *After the Future*. Edinburgh, UK: AK Press, 2011.

Braidotti, Rosi. *The Posthuman*. Oxford: Polity, 2013.

Brynjolfsson, E. and McAfee, A. *The Second Machine Age: Work, progress, and prosperity in a time of brilliant technologies*. New York: W. W. Norton & Company, 2014.

Clark, Andy. *Natural-Born Cyborgs: Minds, Technologies, and the Future of Human Intelligence*. Oxford: Oxford University Press, 2003.

Corrington, Robert S. *The Community of Interpreters: On the Hermeneutics of Nature and the Bible in the American Philosophical Tradition*. Korean translation by Iljoon Park. Seoul: Dong-Yun, 2018. Originally published by Mercer University Press in 1996.

Executive Office of the President & National Science and Technology Council Committee on Technology. *Preparing for the Future of Artificial Intelligence: A Government Report*. Washington DC, White House: Createspace Independent Publishing Platform, 2017.

Harari, Yuval N. *Homo Deus: A Brief History of Tomorrow*. London: Harvill Secker, 2015.

Illich, Ivan. *The Right to Useful Unemployment and Its Professional Enemies*. London: Marion Boyars, 1978.

Kim, Jae-Hee. *Techno-Philosophy of Gilbert Simondon: A Blueprint for the Posthuman Society*. Korean Publication. Paju, SK: Acanet, 2017.

Lee, Kwang-Seok. *Criticizing the Data Society*. Seoul: Chak-Il-Neun-Soo-Yo-Il, 2017.

Lloyd, Genevieve. *Spinoza and the Ethics*. London & New York: Routledge, 1996.

Nussbaum, M. *Frontiers of Justice. Disability, Nationality, Species Membership*. Cambridge, MA: Harvard University Press, 2006.

Sandel, Michael J. *The Case Against Perfection: Ethics in an Age of Genetic Engineering*. Korean Translation by Su-Kyung Lee. Seoul: Wiseberry, 2016.

Sax, David. *The Revenge of Analog: Real Thing and Why They Matter*. New York: Ingram Publisher Services, 2016.

Schoen, M. and K. Loberg. *Your Survival Instinct Is Killing You: Retrain Your Brain to Conquer Fear and Build Resilience*. New York: Plume Books, 2013.

Shin, Sang-Gyu. 2014. *The Future of Homo Sapiens: The Transhuman and the Transhumanism*. Paju: Acanet.

———. "The extended mind and the expansion of the self." in Ewha Institute for the Humanities & Labex Arts-H2H Institute. *The Theater of the Posthuman*. Paju: Acanet, 2015.

Part V

(A)THEODICY THROUGH THE ANTHROPOCENE

Chapter 12

Selving in a Dangerous World

William James, Buddhism, and Ecstatic Naturalism

Jonathan Weidenbaum

INTRODUCTION: RUGGED INDIVIDUALISM

If one were looking for evidence that even the most abstract and rigorous of metaphysical problems can have practical and real-life bearing, consider the following questions pertaining to the self: What precisely is the "I" or "me"? If everything in my body changes over time, from my hair follicles and skin cells down to my neurons, what makes me the same being I am now that I was, say, twenty years ago? Do we possess a consistent identity or transcendental self functioning beneath the cultural influences shaping (or creating?) my language, patterns of behavior, and appearance? What is the proper direction we must take in order to realize and honor the nature of human personhood: the building up and sharpening of our individuality, or the deepening of our sense of belonging to a greater whole? To what extent are these two missions reconcilable?

On more than one occasion, I recall family members priding themselves, in good American fashion, on a certain *rugged individualism*—the capacity of the self to define and direct its own nature, to draw upon its own energies to surmount any and all problems it may encounter. But the notion of the individual self, particularly in its rugged variety, has taken a beating in the recent history of ideas. Indeed, one contemporary psychoanalytical thinker, Roger Frie, derides the classical Euro-American stress upon the individual and autonomous self by equating it with the image of the Marlboro Man, a popular advertising character for cigarettes.[1] For the purposes of championing a more social and culturally conditioned understanding of the self, this is the perfect satirical metaphor. On the one hand, as a strapping and handsome cowboy straight out of a Western, clasping a cigarette between his lips as he stands confidently against the open frontier, the Marlboro Man is *the* emblem of rugged individualism, masculinity, and dominion over the environment. And yet, behind the bracing visuals, this iconic figure really is a tool through which corporations manipulate the masses into purchasing and ingesting poison.

In choosing this illustration, Frie's message is clear. First, the self as envisioned by much of Western civilization—the product of Cartesian and enlightenment doctrines of rationality, self-transparency, individualism and autonomy—is an illusion now quite useful for a consumerist and capitalist society. And second, it is a *lethal* illusion—if unchecked by philosophies which stress our harmony with, rather than mastery over, the environment along with the broader social good. From this perspective, it might be more than symbolic that a number of models for the Marlboro Man have died of smoking-related forms of cancer.[2]

Of course, some may respond that the active self, particularly when employing a good measure of rationality and objectivity, is precisely what brings the perils of cigarettes to light. Frie would not disagree. His project, in fact, is nothing less than to rediscover a measure of the individual and the volitional through a more contextualized understanding of the human condition. In order to enact just such a synthesis, Frie and his coauthors make use of major figures in psychoanalysis and Continental philosophy.

The purpose of this chapter is to explore another reconciliation of the individual and active self with its broader context. And similar to Frie, it does so by drawing upon a philosophical tradition—albeit one far grander and systematic in scope. This is the ecstatic naturalism of Robert Corrington. Corrington's frequent articulation of something called the *Selving process* is suggestive in its very language: on the one hand, we see in it a reference to a self—and yet not the self as a solid and finished entity or substance, but a process, an activity, something underway, in development, unfinished.[3] This activity-entity is both thoroughly semiotic (composed of meanings) and unique in that it straddles two levels of reality: first, it finds itself within this endlessly textured universe of plants and planets, cells and supernovas; and second, it is connected, through its darker and unconscious depths, to the more chaotic and dynamic forces crackling their way through, and incessantly crystallizing into that first, more visible realm.

Like Frie, Corrington draws upon Continental thought and currents in psychoanalysis. But ecstatic naturalism is even more a child of the Transcendentalists and classical American philosophy. In order to unpack Corrington's concept of Selving in all of its metaphysical richness, I therefore enlist one of the main forerunners and influences upon ecstatic naturalism. This is the philosopher and psychologist William James. In what follows,

I first examine the pathbreaking thought of James on personal identity, followed by his take on the underlying ontological field in which subjectivity and agency make their appearance. I then return to further articulate the details of the Selving process as envisioned by Corrington.

In one of his most illuminating treatments of the Selving process, Corrington outlines his agreements and disagreements with the third-century Indian philosopher Nāgārjuna.[4] It is common for those who explore the topic of personal identity to engage with Buddhism, a tradition which advocates the doctrine of *anatta* meaning no soul or eternal self.[5] It is no surprise that there have been quite a few important comparisons of James and Buddhism on personal identity.[6] Buddhist thinkers will therefore play a role throughout this chapter—mainly for the purpose of further disclosing what ecstatic naturalism has to offer our understanding of human existence.

Admittedly, quite a few names and traditions have already been mentioned in this introduction. I make no apologies for this, for part of what makes ecstatic naturalism so compelling is the boldly eclectic and inclusive character of its speculations.

THAT MIXTURE OF UNITY AND DIVERSITY

The Self and *Anatta*

In truth, one can fill an entire essay, even several essays—mining the relevance of James for insights into the nature of the self. But it is only chapter X of the *Principles of Psychology*, "The Consciousness of Self," which focuses explicitly and extensively on this topic.[7] Here I present an exceedingly brief synopsis of a number of its more salient points regarding the question of personal identity—those which will help us illuminate a few things about Corrington's notion of the Selving process.

In *The Principles*, James famously distinguishes between the *me*, or how we appear to ourselves, and the segment of our consciousness to which the me appears—namely the "I." These forms of the *me* include our body and possessions, our standing with others, and our conscience. James labels these as the *material*, *social*, and *spiritual* selves, respectively.[8]

We must choose a self with which to identify over our life span. But since we are finite beings, we cannot entertain more than a limited number of these potential selves. It is the *social* me, in fact, which projects ahead of our current selves a kind of ideal toward which to aspire.[9] This picture of the self as not something ready-made but a project is reminiscent of the existentialists, among others.

At one point, James entertains the possibility of our consciousness as entirely nonpersonal at its base, devoid of an ego whatsoever—what he calls *sciousness*.[10] But consummate phenomenologist that he was, James could not rest content with an account of our conscious life with *no* personal dimension whatsoever. His depiction of the "I" therefore steers between a complete rejection of any personal identity—as represented in David Hume's picture of our interior life as composed entirely of discrete sensory bits with no unifying elements whatsoever—and the traditional picture of a soul-substance or pure unity underlying and continuing through an only apparent diversity.

Against these extreme and mutually exclusive alternatives, James adds to his famous picture of our mental life from the well-known previous chapter—that of a "stream of thought"—with a succession of interrelated pulses of awareness.[11] The "I" is essentially a passing series of thoughts, each one identifying with and appropriating the one before.[12] It is important to note that each thought recognizes (or really fashions) its link with a previous thought based upon the "warmth and intimacy" it feels with it and that shared warmth is based upon the connection of both with our body. It is the *material* self, mentioned above, which serves as a point of reference by which moments of consciousness identify with previous ones.

Against the idea, taken for granted by most of us, that a self somewhere deep within our psyche sends out our thoughts, James memorably asserts that "the passing Thought then seems to be the Thinker."[13] As a succession of thoughts, each of which reaches back to claim the one before, the Jamesean self is neither a preexisting entity (like the traditional notion of the soul) nor a formless multiplicity of discrete elements (as in classical empiricism), but "that mixture of unity and diversity"—to quote once more from *The Principles*.[14]

Two core endeavors are found throughout James's authorship. There is, first, the attempt to do justice to the full texture of our actual experience—in contrast to the influence of a few outdated philosophical models of it. Second, James aims to encourage and intensify our ethical engagements with the world, what he calls the "strenuous mood."[15] One convenient means of illuminating the connection between these two goals is to examine how a few authors employ the thought of James in order to disclose the true subtlety of the Buddha on the question of personal identity.

It is very tempting to equate the Buddha's doctrine of *anatta*—the rejection of a permanent and unchanging self—with Hume's well-known skepticism concerning the presence of an "I" abiding within our sense experiences.[16] A few scholars of Buddhism, however, have forcefully argued that the thought of James makes for a far better comparison to the psychology of the Buddha than Hume's empiricism.[17] Earlier, we saw James's avoidance of both Hume's complete denial of the self on the one hand and the traditional soul-substance idea on the other. Similarly, the Buddha sought a middle way between the annhilationists of his time who rejected *any* sense of identity whatsoever and the Brahmanical tradition which advocated an eternal and transcendental self.[18] Although rejecting the *atman* or eternal self of Hinduism, the Buddha sought to account for the features of our subjectivity which, among other things, make the ethical life possible. Hume's depiction of our inner life "in terms of passive particulars," argues D.C. Mathur, "had no intelligible explanation for the facts of initiative and moral effort."[19] Without a means of carrying over our memories from one mental state to another, for example, there can be no sense of personal responsibility, no meaningful break from the past. In order to provide for this sense of continuity without falling back upon a soul or transcendental self, the Buddha depicted the psyche as an interrelated flow of experiences rather than a collection of sensory bits, a *viññā ṇa-sota* or "stream of consciousness."[20] The similarity of this concept with James's stream of thought, a concept created more than two millennia later, is almost uncanny.

Pure Experience and Sunyata

"The parallels between the Buddha's non-substantialist philosophy and that of William James are so many," states David Kalupahana, "that one begins to wonder about the extent to which the later may have been influenced by the former."[21] James's deconstructing of the self in *The Principles*, albeit with his similar denouncing of the Humean dismissal of personal identity, is not the only parallel between the thought of James and Buddhism. If *The Principles* depicts our conscious life as a fluid and overlapping succession of thoughts, James's mature metaphysics broadens this picture to encompass all that exists. In the later thought of James, both mind and matter and subject and object are but functional expressions woven out of a more fundamental reality called *pure experience*.[22] The latter is not to be understood as a single and identifiable substratum akin to most previous forms of monism (i.e., Thales's claim that all is water). It is pure immediacy: "a simple *that*" which, interpreted from one angle is a conscious act of perception, and articulated from another angle is a piece of matter that is perceived (emphasis in original).[23] James's concept of the self is not too different between his psychology and his later metaphysics: in both cases it is not some independent and immutable entity, but an operation or point of reference occurring within and through a dynamic field of relations.[24]

Similar to the shift of focus between the early and later James, Mahayana Buddhism extends the denial of a permanent and substantial identity beyond the confines of our interior life and applies it toward the entire cosmos. Perceiving that all things are causally and ontologically dependent upon everything else, the Mahayana tradition generally holds that nothing possesses a strictly independent existence.[25] In short, all is part of the sunyata:

the void, or emptiness (we will see more of this concept below). It should therefore be of little surprise that the founder of a more recent movement of Buddhist-inspired philosophers would find James's concept of pure experience to be of great use for articulating his metaphysical vision.[26] And yet, a comparison of James with Buddhist thought is equally illuminating in their differences. For James remained a pluralist to the end, and argued that the disconnections between the events comprising the cosmos are just as real as their connections.[27]

Acknowledging the presence of rifts and discordances within the cosmos connects with the two aforementioned aims of James's thought: namely, capturing the full range of our concrete experience and encouraging the ethical life. In regard to the former, an adequate philosophy must account for the rougher and grittier features of existence. In one of my favorite of his quotes, James likens the perfectly unified and holistic metaphysics so popular among the idealists of his time to an uptight and pretentious clergyman: "It seems too buttoned-up and clean-shaven a thing to speak for the vast slow-breathing unconscious Kosmos with its dread abysses and its unknown tides."[28] This messier and only roughly connected world—what James calls a *concatenated* universe—also supports the morally strenuous mood.[29] In *Pragmatism*, James opposes both a pessimistic and an optimistic picture of the cosmos—situations where the world's salvation is either impossible or guaranteed, respectively—for a *melioristic* one in which individual and collective effort can make a difference.[30] In "Is Life Worth Living?" James rousingly affirms: "For such a half-wild, half-saved universe our nature is adapted."[31] In the interest of capturing the precariousness of life and promoting personal exertion, James argues for a finite God, a being for whom we are not passive subjects but partners in improving the cosmos.[32]

For our purposes here, it is necessary to linger just a little more on the religious element in the Jamesean self. James's finite deity is not a transcendent being, one outside or above the world, but a greater consciousness in which our particularized selves are ensconced.[33] Our fundamental access to this deeper reality is primarily downward through the recesses of our psyche, at levels often far beneath our surface awareness. The divine touches us through the less accessible borders of our minds, most typically when we are at our most desperate and our own efforts for self-healing have collapsed.[34]

To summarize: the Jamesean self is not a thing but a process. In *The Principles*, it is an activity in perpetual self-constitution. In the later metaphysics of James, it exists as part of a broader web of relations. Under certain conditions, its depths may open to the transformative powers of what lies beneath it. Finally, the cosmos of which the self is an integral part is neither a fully interconnected whole nor a hard and atomic pluralism. This is a concatenated universe, a metaphysical enlarging of James's account of our inner life in *The Principles*—namely, an uneasy mixture of unity and diversity.

MANY ONES AND MANY PLURALITIES

The Selving Process

As I have mentioned in the introduction, human beings have a unique place in the metaphysics of ecstatic naturalism. On the one hand, our body and much of our personality are woven out of the same material we perceive receding away from us in all directions. On the other hand, the deeper and unconscious layers of our psyche reach downward into the turbulent and formless depths continually bursting into, and shaping, the more apparent world we commonly perceive with our senses. Under the influence of a line of thought stretching from the Scholastics through Spinoza and onward to the present, Corrington labels these two sides of reality as *nature natured* and *nature naturing*, respectively.[35] As spiritual and aesthetic meanings spark upward to reverberate through our psyche, occasionally imbuing specific features of our environment with something of the numinous (those overlapping layers of significance, or "semiotic density," which lead to the "sacred folds" in nature), so we are presented here with a theology that rejects the supernatural—a proper ecstatic *naturalism*.[36]

Corrington distinguishes between two types of culture: the *natural community* is an authoritarian society marked by, among other things, rigidly interpreted dogmas, enforced social roles, and repression of much that is vital in the human being. The *community of interpreters*, by contrast, is open to novel insights and forms of meaning, a softening and reinterpretation of social roles, and the release rather than repression of our healthier instincts.[37] Against the tendency of many cultures toward strong forms of social control and the repetition of the past, we can appreciate one of Corrington's most succinct and direct formulations of the final aim of the human self: "The ultimate goal of the Selving process is precisely to increase the amount of free will and creativity in the host self and tribe." In ecstatic naturalism, it is the inspiration of the saint and the artist where purpose and direction for humanity are found: "Here, and only here, we can safely introduce the notion of teleology into the system."[38] The receptivity of the human being to nature naturing, along with the cultivation and protection accorded those so gifted (or cursed) with such a capacity, therefore possesses as much a social as a spiritual value.[39]

The relationship of the self to semiotics in ecstatic naturalism expresses a reconciliation between the value of the individual and an acknowledgment of its rooting within both nature and culture. "The human process," explains Corrington, "is fully embedded in infinite streams of semiosis that reach down to all sentient beings, from the archetypes to plants and perhaps beyond."[40] But it is the creative self—one attuned to the spontaneous heart of nature naturing—who participates in the reinterpretation of old meanings and the fashioning of new ones. Here we see the full significance of Corrington's label for progressive societies as *interpretative* communities (an influence from the philosopher and chief semiotician, Charles Peirce). And as with semiotics, so with the arts. Through a rapport with nature naturing, the genius creates not only beautiful things by which to advance civilization. A true artist also shapes and refines what is of equal importance for furthering the community of interpreters: their own unique self.[41]

We have acknowledged the route taken by James in *The Principles* between the unchanging soul-substance and Hume's analysis of our conscious life into a series of sense particulars. In an engagement with Buddhist thought in the final chapter in *Nature and Nothingness*, Corrington articulates a similar middle ground for the Selving process. In ecstatic naturalism, the self is neither the complete illusion and product of vanity that Corrington's interpretation of Buddhism takes it to be nor the unchanging and substantial self of some Eastern and Western belief.[42]

Of course, we have seen that Buddhist reflections on the self—particularly as disclosed by scholars of early Buddhism—is *itself* a middle way between an absence of anything personal in the psyche and the eternal soul of the Vedic tradition. The motivation of early Buddhism and the thought of James in offering a richer and more nuanced psychology is to better provide for the volitional and ethical life, as well as other forms of human fulfillment. Ecstatic naturalisms's take on the Selving process is fashioned along similar lines.

Corrington's remarks about Buddhism are found in the context of his engagement with Nāgārjuna, a third-century Indian philosopher who pioneered the Madhyamaka or "middle way" school of Mahayana Buddhism as well as the notion of the sunyata or "void." Nāgārjuna's ideas are addressed by Corrington as a means of exploring *encompassing nothingness*—the most primordial and conceptually elusive kind of nothingness in the metaphysics of ecstatic naturalism.[43] Encompassing nothingness may best be defined as an unconditioned reality pervading and giving birth to *both* nature naturing and nature natured. Corrington describes it as the "nongrounded ground" of all forms in the cosmos.[44]

A few general remarks on the thought of Nāgārjuna are relevant here. Against earlier schools of Buddhism and other Indian thinkers of his time, the main figure of the "middle way" argues that all phenomena are empty of a substantial or independent essence (*svabhāva*).[45] His most revered text, the *Mūlamadhyamakakārikā* (I am using the same translation as Corrington), employs rigorous conceptual arguments in order to demonstrate, paradoxically, the futility of all concepts. Nāgārjuna argues that in the face of the interdependent and ineffable nature of all existence, not only do our attempts to philosophically justify commonsense structures like causality, motion, and perception fail—but even the notion of emptiness is *itself* empty.[46] Nāgārjuna's plan is therefore *not* one of clearing away all metaphysics of substance only to reveal yet another fundamental category by which to grasp the cosmos—as if nothingness were a self-subsisting and transcendental reality like God or *brahman*.[47] It is to reveal the complete impossibility of any sort of metaphysics to capture the ultimate nature of things. Free of all attachments to concepts whatsoever, we may then realize the emptiness of everything in a direct and experiential rather than a merely cognitive manner. Like the Buddha himself, Nāgārjuna's final aim for human beings is our spiritual liberation.

The goal of the rest of this section is not to evaluate the accuracy of Corrington's interpretation and critique of Nāgārjuna—although I do address one place where the former's take on Buddhism falls short. I seek instead to employ Corrington's dialogue with the great South Asian thinker as a means of disclosing what is uniquely relevant in ecstatic naturalism, particularly in regard to the Selving process. Here, all of our pieces fit together, as the themes from the preceding sections come to a head. Perceiving the Selving process in the light of William James, as well as in contrast with Buddhism, may assist us in appreciating what is truly distinctive in ecstatic naturalism's interpretation of human purpose. But first, we require an acquaintance with one more figure.

Many Goals

A particularly strong influence on ecstatic naturalism is the underappreciated American philosopher Justus Buchler, whose ontology and method alike are crucial for a complete understanding of Corringtron's system. For our somewhat more limited task, it is enough to recognize just a few Buchlerian themes. One of these is the idea of the universe as composed of a limitless number of *natural complexes*.[48] As the essential unit of Buchler's (and Corrington's) metaphysics, this is best understood as an integral structure of relations rather than any particular substance. Hence, thoughts, animals, gas clouds, bricks, fantasies, novels—indeed, anything—is a complex.[49] Moreover, as the constituents of any complex are itself complex, so each and every complex is itself a trait within another complex—a nesting without beginning and end. Even the *relations* between complexes are themselves complexes. It is only in reference to a particular context or angle of inquiry that any complex can be distinguished from the others.

So far, the kinship between the thought of Corrington and Buchler with the thoroughly interdependent nature of the cosmos in Mahayana Buddhism appears very strong. But there are two essential differences. The first is one of purpose. The thought of Corrington/Buchler seeks to intensify rather than degrade the metaphysical quest, and hungrily exploits the opportunity for a ceaseless investigation of each and every complex in regard to their endless meanings, relations, and so on. While, as we have seen, the thought of Nāgārjuna encourages a *release* from all conceptualizing and metaphysical generalization for the goal of spiritual liberation.[50] A second, and more crucial difference for our interests, is an ontological feature we have already seen in James. Buchler insists that while every natural complex is related to another complex, *none* is related to every complex.[51] Thus the Jamesean picture of the disconnections between the events comprising the cosmos finds it analogue (or its influence?) in Buchler and Corrington. Placing both of these differences together: to Nāgārjuna's rejection of both "singularity" and "plurality" in the cosmos—part of his insistence on the futility of all conceptual contrasts—Corrington finds that "the language of William James is the more appealing; namely there are many ones and many pluralities."[52]

In his *Essays in Radical Empiricism*, James argues that things are, finally, not reducible to a web of relations, and that individual facts may defy the "through and through" cosmos of the monists and idealists of his time.[53] A concatenated universe of only partially overlapping phenomena will more readily yield coherent purposes as well as objects. Corrington quotes Nāgārjuna on how "the nectar of the teachings of the buddhas" is "not having a single goal, not having many goals, not destroyed, not eternal."[54] To these spiritual instructions, Corrington replies that Nāgārjuna may be correct about there being no overriding goal in the cosmos, but the Buddhist philosopher is mistaken on the second part: for there are, in fact, as many little teleologies in nature as there are human beings. "The Selving process creates and sustains many goals during its trajectory through time," Corrington writes, "and some of them may last a lifetime and bear fruitful results for future generations."[55]

Mahayana Buddhism should never be thought of as completely evaporating the self into some indiscriminate and universal vacuity. At one point, Corrington states that "unlike most, if not all, forms of Buddhism, Selving requires a centered personality that is formed under the principle of individuation." He recognizes that the personal elements within the psyche "remain at least partially intact even with the entrance of nothingness."[56] In Mahayana Buddhist literature, however, the enlightened personality is similarly enriched after *satori* or enlightenment, and the individual is no more nullified here than in early Buddhist scriptures (discussed in the previous section).[57] With that stated, the particular intensity and sense of direction granted the Selving process are far more sharply defined in ecstatic naturalism.

An essential theme of *Nature and Nothingness* is how the harsher and less savory parts of the universe—the loss of meaning we suffer in our everyday lives, the complete indifference of the universe to all sentient beings, and the occasionally hazardous energies pulsating through the natural world—may pierce through our defenses and illusions and serve as catalysts for our self-knowledge and development.[58] For instance, Corrington outlines how the right approach to anxiety, death, and other forms of trauma, what he calls *holes in nature*, can help break away our "emotional armoring" and neurotic attachments.[59] Thinking over these forms of nothingness, my attention is drawn once more to James's darkly picturesque description of nature as "a vast slow-breathing unconscious Kosmos with its dread abysses and its unknown tides."[60] We have also seen that the highest task of the Selving process, in ecstatic naturalism, is to serve as a bulwark against the tendencies exploited by the natural community, including our all-to-human retreat into totalitarian control, inertia, and so on. In regard to our responsibilities in the face of this precarious and unfashioned universe, that which includes the dangers inherent within the human community, I am also reminded of James's encouragement to sharpen the morally "strenuous mood."[61]

There is no easy, one-to-one match up between James and Corrington on matters of personal identity or any other topic pertaining to human existence. First, ecstatic naturalism fills out the Selving process with an abundance of insights from psychoanalytical literature, semiotics, and Continental philosophy. Second, Corrington's depiction of the world runs a few shades darker than the melioristic universe of James: the future extinction of our species, for instance, is a much stronger possibility for the former.[62] Third, James's *pluralistic pantheism*—that greater but finite mind underlying and occasionally influencing our smaller selves—may be the ancestor to Corrington's deep pantheism in numerous ways. Yet the deity of James is still a personal entity, something unwarranted from the nontheistic perspective of ecstatic naturalism.[63]

In the end, the basic similarities between James and ecstatic naturalism, along with Buddhism, is likely of more value than their differences. Let us distill their insights: the self is a process instead of a thing; it is neither nonexistent nor an unchanging entity or substance. It exists in and through a larger field of relations, and yet possesses an integrity of its own. Buddhism in all of its varieties and denominations has a rich ethics, but what is distinctive to James and Corrington is the recognition of those breaks and rifts in existence, along with the dangers inherent to the human community, which assist in granting the self its larger purpose. Here is the reconciliation promised in the introduction: namely, between the robustly active individual and its inherence within a larger context.

CONCLUSION: INTIMACY

It can be argued that James, Corrington, and Buddhism together can offer an alternative take on the Marlboro Man. With his brawny appearance set against an equally hardy landscape, the relationship of the Marlboro Man to nature need not be understood as one of privilege or domination. Seen in the light of our two American philosophers and one wisdom tradition, the famous cowboy may reflect what James in *A Pluralistic Universe* calls *intimacy*: a kind of organic continuity between the divine, the world, and ourselves.[64] With his wind and sand-beaten countenance, the Marlboro Man appears more as a living outgrowth of his environment than an imposition from the outside; a slightly rougher take on the sage of the Asian landscape paintings, whose quiet steps through woods and mountainsides leave barely a footprint.[65] But our hero can, and should, make his mark on things. Perhaps it is through harmonizing the discordances within our surroundings that we learn to refine our own harsher edges, and to replace rugged forms of Selving with more intimate ones. In this way, freeing himself from his role as an advertising character for cigarettes, even standing as a symbol of their danger, the Marlboro Man may both reshape himself and do his small part for the community of interpreters.

NOTES

1. Roger Frie. "Culture and Context: From Individualism to Situated Experience," in *Persons in Context: The Challenge of Individuality in Theory and Practice*, eds. Roger Frie and William J. Coburn (New York: Routledge, 2011), 5.
2. Matt Pearce, "At least four Marlboro Men have died of smoking-related diseases." Los Angeles Times, January 27, 2014. https://www.latimes.com/nation/nationnow/la-na-nn-marlboro-men-20140127-story.html.
3. While Corrington explores "the Selving process" in most of his major works, I make the most use of his latest work at the time of writing: *Nature and Nothingness*. Robert Corrington. *Nature and Nothingness: An Essay in Ordinal Phenomenology* (New York: Lexington Books, 2017). References throughout will also be made to his prior volume: *Deep Pantheism: Toward a New Transcendentalism* (New York: Lexington Books, 2016).
4. Corrington, *Nature and Nothingness*, 142–148.
5. While the Buddhist concept of the self will be treated repeatedly below, one well-known place to begin exploring the idea of *anatta* is a much beloved introduction to Buddhism: Walpola Rahula, *What the Buddha Taught* (New York: Grove Weidenfeld, 1974). See the sixth chapter.
6. A number of these comparisons are presented below.

7. William James. *The Principles of Psychology, Volume One* (New York: Dover Publications, Inc., 1950). Gerald E. Myers states as follows: "If a topic interested James, he discussed it in essay upon essay; many of the chapters in the *Principles*, for instance, are elaborations of earlier articles or are elaborated in later ones. But the concept of the self, while a profound concern of his, is treated almost exclusively in 'The Consciousness of Self.'" Gerald E. Myers, *William James: His Life and Thought* (New Haven: Yale University Press, 1986), 344.

8. James, *The Principles of Psychology*, 291–330.

9. Ibid., 314–316.

10. Ibid., 304.

11. Ibid, chapter IX.

12. The core of this discussion is ibid., pages 337–342.

13. Ibid., 1950, 342.

14. Ibid, 1950, 352.

15. William James, *The Will to Believe and Other Essays in Popular Philosophy* (New York: Dover Publications, Inc., 1956), 211.

16. In the past, the present author has been guilty of succumbing to this temptation.

17. D. C. Mathur, "The Historical Buddha (Gotama), Hume, and James in the Self: Comparisons and Evaluations." *Philosophy East and West* 28, no. 3 (1978): 253–269. David J. Kalupahana, *The Principles of Buddhist Psychology* (Albany: State University of New York Press, 1987).

18. Mathur, 255–256, and Kalupahana, 40.

19. Mathur, 262.

20. Kalupahana, 21. Kalupahana writes as follows: "The Buddha did not analyse experience into discrete entities or momentary impressions and then try to find a way of unifying them. Such an enterprise was undertaken by his misguided disciples a few centuries later, namely the Sarvāstivādins and the Sautrāntikas." Kalupahana, 22.

21. Ibid., 10.

22. See James's "A World of Pure Experience." William James, *Essays in Radical Empiricism* (Lincoln: University of Nebraska Press, 1996), 39–91.

23. James, *Essays in Radical Empiricism*, 23.

24. Gerald Myers states: "The language changed somewhat—the stream of thought in *Principles* became a field of consciousness or a portion of pure experience—but basically the same doctrine of the self was retained in the later metaphysics." Myers, 353.

25. Corrington states: "If earlier *Theravada* texts tended to limit nothingness to the human process, *Mahayana* texts extended the concept to things and to the universe." Corrington, *Nature and Nothingness*, 140.

26. Nobuo Kazashi offers summary of the impact of James's pure experience on Kitaro Nishida, the founder of the Kyoto school of Buddhist philosophy. Nobuo Kazashi, "Metamorphoses of 'Pure Experience': Buddhist, Enactive and Historical Turns in Nishida." In *Education and the Kyoto School of Philosophy*, eds. P. Standish and N. Saito (New York: Springer, 2012), 77–89.

27. James, *Essays in Radical Empiricism*, 44–52. Kazashi includes James's account of the disjunction between experiences as one of the two differences between the American philosopher and the Buddhist thought of Nishida. The other is that the "state of 'pure experience' signifies, in Nishida's version, not only the originary state of experience which is 'not yet' divided into the subjective and the objective, but also, and more importantly, the 'ideal' state of experience." Kazashi, "Metamorphoses of 'Pure Experience'," 82.

28. James, *Essays in Radical Empiricism*, 277–278.

29. For one place where he introduces the concept of a *concatenated* universe, see: James, *Essays in Radical Empiricism*, 107.

30. James, *Pragmatism*, 137.

31. James, *The Will to Believe*, 61.

32. James argues for a finite deity in many of his writings, but his most completed statement on the matter is found here: William James, *A Pluralistic Universe* (Lincoln: University of Nebraska Press, 1996), Lectures I and VIII.

33. Ibid., lectures I and VIII. William James, *The Varieties of Religious Experience* (New York: Barnes & Nobles Classics, 2004), lectures IX and "Conclusions."

34. James, *Varieties of Religious Experience*, lectures IX and "Conclusions."

35. Corrington, *Nature and Nothingness*, ix. Also: *Deep Pantheism*, chapter 1.

36. Corrington, *Deep Pantheism*, 42.

37. Corrington, *Nature and Nothingness*, ix.

38. Ibid., 66.

39. Corrington has done much work on the relationship between bipolar disorder, genius, and the Selving process. For a harrowing personal account of this, along with a study of the topic in general, see: Robert Corrington, *Riding the Windhorse: Manic-Depressive Disorder and the Quest for Wholeness* (Latham: Hamilton Books, 2003).

40. Corrington, *Nature and Nothingness*, 55.

41. Ibid., 34, 77. Corrington here draws upon the work of Otto Rank.

42. In his own words: "The ordinal position is thus midway between a total nirvana with its nonself and a clumsy ham fisted belief in an unchanging brute substance that may take on attributes, but doesn't itself change." Ibid., 147–148.

43. Ibid., xi–xii, chapter 6.

44. Ibid., 136.

45. Nāgārjuna, *Middle Way*, Mark Siderits and Shōryū Katsura, eds., (Somerville: Wisdom Publications, 2013), 1.

46. Ibid., chapter 24.

47. One of the most lucid interpretations of Nāgārjuna's thought, including his avoidance of reifying emptiness into something metaphysical, is found here: Chakravarthi Ram-Prasad, *Eastern Philosophy* (London: Weidenfeld & Nicolson, 2005), 30–33, 130, 171–173, 193–197, 233–234.

48. Justus Buchler, *Metaphysics of Natural Complexes*, Kathleen Wallace, Armen Marsoobian, and Robert Corrington, eds. (Albany: State University of New York Press, 1990).

49. Ibid., chapter 1, "Rudimentary Considerations."

50. For a keen comparison between Mahayana Buddhism and the thought of Justus Buchler, see: Marjorie Cantor Miller, "The Concept of Identity in Justus Buchler and

Mahayana Buddhism." In *Nature's Perspectives: Prospects for Ordinal Metaphysics*, Armen Marsoobian, Kathleen Wallace, and Robert S. Corrington, eds. (Albany: State University of New York Press, 1991), 93–113. On the main difference between Mahayana Buddhism and Buchler, Miller writes: "For Buchler, precise identification is a tool leading to ever more fruitful metaphysical investigation; for the Buddhist, the very similarly conceived process of identification is ultimately counterproductive in the achievement of the ideal spiritual mode of human being-in-the-world. Metaphysics itself is ultimately transcended; in the enlightened state it is finally irrelevant." Miller, "The Concept of Identity," 111–112.

51. Buchler, *Metaphysics of Natural Complexes*, 24.

52. Corrington, *Nature and Nothingness*, 141. In regard to Nāgārjuna, the original is found in the opening verses of the Mūlamadhyamakakārikā, as part of the "Dedicatory Verse" to "The Fully Enlightened One." Nāgārjuna, Mūlamadhyamakakārikā, 13. In regard to James: *Pragmatism*, 73.

53. James, *Essays in Radical Empiricism*, 266–279.

54. Corrington, *Nature and Nothingness*, 144. Original: Nāgārjuna, *Middle Way*, 205.

55. Corrington, *Nature and Nothingness*, 144.

56. Ibid., 147.

57. A major example of this is the tenth and final panel in the Zen Oxherding Pictures: a set of images representing the stages of the mind on its journey toward enlightenment. While the eighth panel is an empty white circle representing a non-dual awareness, this image does not have the last word. The tenth and final picture is that of a smiling and joyous monk (named Hotai) strolling through the marketplace, spreading his love to everyone he sees. According to D. T. Suzuki, the most well-known version of the ten Oxherding Pictures is attributed to a Zen master named Kaku-an Shi-en. Daisetz Teitaro Suzuki, *Manuel of Zen Buddhism*, New York, Grove Weidenfeld, 1960, 127–144. That Corrington recognizes how the most primordial sort of nothingness does not nullify finite entities is evident in the final chapter of *Nature and Nothingness*. See his example of an Asian landscape painting in which "encompassing nothingness shines through in its distinctive kind of radiance." Corrington, *Nature and Nothingness*, 146–147.

58. These first three forms of nothingness are labeled by Corrington as "holes in nature," "totalizing nothingness," and "naturing nothingness," respectively. Corrington, *Nature and Nothingness*, ix–xiii, and chapters 1 through 3.

59. Ibid., 8–12. The idea of emotional armoring is strongly influenced by Wilhelm Reich.

60. See note 25 above.

61. See note 13 above.

62. For instance, see: Corrington, *Nature and Nothingness*, 36.

63. There is therefore a place in the thought of James for a teleology or purpose beyond our human selves. James, *Varieties of Religious Experience*, Conclusions and Postscript; *A Pluralistic Universe*, Lectures I and VIII.

64. James, *A Pluralistic Universe*, Lecture I.

65. Jonathan Weidenbaum would like to thank Guy Woodward for his generous assistance with numerous drafts of this chapter.

REFERENCES

Buchler, Justus. *Metaphysics of Natural Complexes*, Second, Expanded Edition. Albany, NY: State University of New York Press, 1990.

Corrington, Robert S. *Deep Pantheism: Toward a New Transcendentalism*. Lanham, MD: Lexington Books, 2016.

———. *Nature and Nothingness: An Essay in Ordinal Phenomenology*. Lanham, MD: Lexington Books, 2017.

———. *Riding the Windhorse: Manic-Depressive Disorder and the Quest for Wholeness*. Latham, MD: Hamilton Books, 2003.

Frie, Robert. "Culture and Context: From Individualism to Situated Experience," in *Persons in Context: The Challenge of Individuality in Theory and Practice*, eds. Roger Frie and William J. Coburn, 5. New York, NY: Routledge, 2011.

James, William. *A Pluralistic Universe*. Lincoln, NE: University of Nebraska Press, 1996.

———. *Essays in Radical Empiricism*. Lincoln, NE: University of Nebraska Press, 1996.

———. *Pragmatism and The Meaning of Truth*. Cambridge, MA: Harvard University Press, 1975.

———. *The Principles of Psychology, Volume One*. New York, NY: Dover Publications, Inc., 1950.

———. *The Varieties of Religious Experience*. New York, NY: Barnes & Nobles Classics, 2004.

———. *The Will to Believe and Other Essays in Popular Philosophy*. Mineola, NY: Dover Publications, Inc., 1956.

Kalupahana, David J. *The Principles of Buddhist Psychology*. Albany, NY: State University of New York Press, 1987.

Kazashi, Nobuo. "Metamorphoses of 'Pure Experience': Buddhist, Enactive and Historical Turns in Nishida," in *Education and the Kyoto School of Philosophy*, eds. P. Standish and N. Saito, 77–90. New York, NY: Springer, 2012.

Mathur, D. C. "The Historical Buddha (Gotama), Hume, and James in the Self: Comparisons and Evaluations," in *Philosophy East and West*, Vol. 28, No. 3, (1978): 253–269.

Miller, Marjorie Cantor. "The Concept of Identity in Justus Buchler and Mahayana Buddhism," in *Nature's Perspectives: Prospects for Ordinal Metaphysics*, eds. Armen Marsoobian, Kathleen Wallace, and Robert S. Corrington, 93–113. Albany, NY: State University of New York Press, 1991.

Myers, Gerald E. *William James: His Life and Thought*. New Haven, CT: Yale University Press, 1986.

Nāgārjuna. *Middle Way*, eds. and trans. Mark Siderits and Shōryū Katsura. Somerville, MA: Wisdom Publications, 2013.

Pearce, Matt. "At least four Marlboro Men Have Died of Smoking-Related Diseases," El Segundo, CA: *Los Angeles Times*, January 27, 2014. https://www.latimes.com/nation/nationnow/la-na-nn-marlboro-men-20140127-story.html.

Rahula, Walpola. *What the Buddha Taught*. New York, NY: Grove Weidenfeld, 1974.

Ram-Prasad, Chakravarthi. *Eastern Philosophy*. London: Weidenfeld & Nicolson, 2005.

Suzuki, Daisetz Teitaro. *Manuel of Zen Buddhism*. New York, NY: Grove Weidenfeld, 1960.

Chapter 13

Redemptive Suffering with *Tianming* 天命

An Ecstatically Naturalist Reading of Sacred Selving in Confucian Ethics

Joseph E. Harroff

INTRODUCTION

This chapter focuses upon some family resemblances of ethical-political thinking between Robert Corrington's ecstatic naturalism as a (post)theological stance and Confucian (*Rujia* 儒家) philosophical discourse of role ethics (倫理道德). Drawing from the philosophy of Mengzi in the Warring States Period (475–221 BCE), I maintain that Corrington's ordinal phenomenology and neo-pragmatism engage in theorizing holes in nature as ecstatically natural complexes and can lead us towards optimizing interfaith conversations. Simply put, Confucian role ethical discourse regarding the very possibility of redemptive suffering and Corrington's appeal to immanent archetypes of non-perfectionist healing and characterological de-armoring can provide theoretical sustenance and ethical hope for those unwilling or incapable of believing in transcendental perfection or ontological transcendence. Such intercultural conversations can prove to be a balm in raising vital questions regarding the morally transformative significance of responsibly contextualizing senseless suffering, moral tragedy, and radical evil within historical-cultural frameworks of interpretation. Such contextualizing doesn't require the positing of any theo-ontological Absolutes or any other forms of non-naturalistic transcendence as supposed omnibenevolent forces in nature, yet it can still vitally energize projects of moral amelioration and political transformation.

TRANSLINGUAL TUNINGS

It is indeed a rare and most appreciated disclosure in the interpretive activity of engaging in comparative philosophical research, to experience edifying ethical and hermeneutic resonance between variegated and diverse semiotic trajectories of distinctive thinking styles and discursive idioms. Separated by the vastly divergent temporal and cultural differences, the Warring States Period the Warring States Period (475–221 BCE) Confucian theorizing of persons as relationally constituted and role-focused human becomings and Robert Corrington's ecstatically naturalistic thinking of persons as always provisional selvings-in-process presents just so many profound resonances for thinking through ethical subjectivity in intercultural conversation.

While researching Tillichian "ultimate concern" and Heideggerian "disclosure" theories of truth as transformative *aletheia* with the interest of developing a responsible comparative methodology for contextualizing issues of religious transcendence, theories of truth, and ultimate values, I first came across Robert Corrington's path-blazing approach to philosophical theology and his many adventurous (post)metaphysical speculations. I increasingly feel that these issues require a nuanced and careful treatment as a kind of pre-theoretic "tuning" for any projects that involve philosophical translating, which is always a matter of translingual practice rather than mere semantic transference.[1] What I find particularly edifying in Corrington's thinking is the framing of a horizonal hermeneutics of an intergenerational, contingent, and future-oriented communicating community of interpretation that, as method, can realize an emergently shared vocabulary for situating intercultural conversations about basic notions of ethical agency, subjectivity, and personhood as sites of sacred and reverential values. The foregrounding of a parity of natural complexes and the robust methodological pluralism inherent in ecstatic naturalism can help in overcoming any historical inertia that would privilege one or another particular trait as being absolutely significant in marking either-or forms of belonging in any given "natural" community. Such exteriorizing and alienating "us and them" or the "West vs. the rest" mentalities always involves massive deployments of characterological armoring (cf. Wilhelm Reich) and the valorization of founding acts of violence via the not so hushed reverberations of such violence echoing into the tragic present. Intercultural comparative philosophizing at its best can help unsettle these oppressive inertias and make space for future-oriented tasks of creative democracy as communicating communities of interpretation (cf. Robert Corrington) and also keep space open for intergenerational communities of inquiry into ultimate truth as a regulative ideal (cf. C.S. Peirce). Interested in promoting such intercultural and translingual spaces required for responsible conversational co-presencing and mutual amelioration, it is important that we take seriously the ethical and political horizons made possible by a reading of *ecstatic naturalism* as a comprehensive horizonal hermeneutic. Such horizons make possible the double-movement of *unsettling* smug subjectivity formations stigmatically marked by fascist, or autocratic, characterological armorings; and *transforming* subjectivities realized in conversational co-presencing with the regulative ideals and affective registers of radical hope for the creative democratic potentials of communicating communities. This is also an ideal horizon for approaching the philosophical translation of certain early Confucian texts emerging from a strikingly different relational ontology of persons and their valuable worlds. Simply put, the need to theorize persons as through-and-through relationally constituted without remainder, while simultaneously being narratively projected in their unique contexts of embodied roles and relationships with family experience serving as an affective center and ongoing imaginative resource, requires a conception of *individuation* that holds natural processes and ecstatic eventfulness in metaxological rhythms of harmonious tension.[2] The very idea of an "intimate universal" ecstatically realized in immanent transcendence resonates quite strongly with the following passage in the *Mengzi*:

> Mencius said, "Those fully exhausting their heartminding, can know their natures (*xing*性). Knowing nature is knowing Heaven (*tian*天). Preserving the [moral] heartmind and nurturing nature—this is to serve Heaven. Premature death or longevity are not the main issues, what really matters is cultivating one's lived-body person (*xiushen*修身). This is how you establish your life as moral vocation (*ming*命)."[3]

Whatever the *Mengzi* may be referencing here by human "nature" is definitely not an abstract universal. And while it may be relatively easy to theorize ethical transcendence relying upon substance ontologies and Absolutist teleological frameworks, when the "blooming buzzing confusion" of lived experience as an infinite array of natural complexes all ordinally understood in a parity of non-Absolutism and hermeneutic pluralism is faced up to, a more nuanced and creative approach to theorizing persons and their valuable worlds emerges as a vital end-in-view.

With questions like these in mind, here I read Corrington in conversation with certain Chinese philosophical sources and seek to creatively deploy a sensitive theoretical imagination and encompassing structure of translingual feeling that facilitates edifyingly transformative intercultural philosophical conversations. With *ecstatic naturalism* as a broadly construed horizonal hermeneutic stance, it should be readily evident that early Confucian articulations of ethical agency resonate quite well with more contemporary relational ontologies of human becomings as unique insistent particularities of body-heart-minding in complex personal-familial-political affairs all happening under the sacred canopy of "all-under-Heaven" (*tianxia*天下). Since these non-foundational and deeply relational interpretive horizons emerge from non-monotheistic, if not atheistic, assumptions seeking to cultivate profoundly sacralizing powers in a pluriversal cosmos, it stands to reason that responsibly contextualized interfaith conversations might help to bring about more ethically dynamic fusions of culturally distinctive interpretive horizons. Such a hermeneutic fusion can help find dynamic beginnings for theorizing anew in our fragile embodiment and increasing relational precarity in the current ensemble of global-scale predicaments like anthropogenic climate chaos, nuclear arms proliferation, imperialism, military-industrial complexes and the rapid expansion and intensification of border-surveillance "security" complexes (to name just a few pressing global-scale predicaments). While always taking a plurisingular process of *becoming individuality* to be a basic unit for thinking, rather than appealing to some ontologically discrete and dogmatically given foundational modicum of *being* as basis for libertarian or neoliberal ideologies of individualism, we can think more relationally about real prospects for healing cultures and future-oriented liberation of uniquely located human potentials for realizing convivial flourishing.[4] Ecstatic naturalism offers profound insights regarding ethical-political agency as creative entanglements of *plurisingularity* eventfully becoming within hermeneutic horizons of deep pantheism, aesthetic holism, and conversational solidarity as *democracy-to-come*. The hope here is that such intercultural comparative philosophical conversations may continue to engender fecund dialogical spaces for translingual transformation between Confucian role ethical thinking about ethical subjectivities rooted in family-born, relationally co-constitutive experience and the ecstatically naturalistic "sacred selving" process. For the Confucians being involved in realizing ethical singularity (*tianrenheyi* 天人合一) is always a matter of co-creativity in an always provisional and open-ended order of ritual constructivism.[5]

PRAGMATISM AND NON-FOUNDATIONAL THEODICY

In interpretively circulating within Corrington's hermeneutic horizon of ecstatic naturalism, I have often found myself drawing from an always evolving philosophical project of semiotic intercultural density and involving the philosophical reconstruction of historical conceptual clusters from classical Chinese thinking. In this chapter I will be especially focusing upon Si-Meng Confucian discourse in articulating a role ethical theoretical framework for thinking through ethical subjectivity.[6] Part of my aim is to be foregrounding various family resemblances emerging between Corrington's ecstatic naturalism as a pragmatic-oriented, anti-metaphysical metaphysics, or humanistic theological stance—*ecstatic naturalism* means much the same thing while sounding much less paradoxical—and the Si-Meng Confucian theorizing of persons as relationally constituted human becomings. We can avail ourselves of Corrington's ordinal phenomenology of natural complexes and his philosophical theology of ecstatic naturalism to be more responsibly approaching the unique historical context and uncommon philosophical assumptions animating a Confucian role ethical discourse.

The task of articulating an edifying "metaphysical" *vision* (an open-ended aspirational view from somewhere, and not some supposedly closed, perfectly complete, and abstract *system*) that can do justice to the more obviously sacred dimensions of our transactional experience, as part of what we might call our unique experience as non-fungible human "becomings"[7] or "sacred selvings," without succumbing to a life-negating, nature-denying and body-denigrating supernaturalist ontology of persons is fraught with a vast array of interpretive and ethical difficulties. Moreover, from the perspective of many postmodern secular thinkers, the adventurous, because risky and difficult, task of articulating a coherent and compelling "naturalistic metaphysics" or "deep theology" of ecstatic naturalism would certainly be considered a thankless one and might even be viewed as pointless or pernicious. Indeed, from the standpoint of certain modern, secular perspectives, it is far from readily obvious what we might hope to gain through such momentous intellectual undertakings. For if we truly embrace our fallible and finite human conditionality, why not just rest content upon our post-linguistic turn, neo-pragmatist laurels and profess, as a viable common faith, some sort of "romantic polytheism" without all the speculative metaphysics, psychoanalysis and ordinal phenomenology as superfluous baggage?[8]

Here I would like to offer a possible response that we could offer such a neo-pragmatist interlocutor, were she skeptical of the worth of naturalistic metaphysical speculation generally and of the merits of an ecstatically naturalistic perspectives in particular. My suggestion is that we might provide a compelling justification in terms of the potential that an ecstatically naturalist perspective has to vitalize any discursive field of theodicy without having to appeal to supernaturalist or ontologically transcendent *θεός*. Indeed, I would argue that it is precisely the capacity that such creative metaphysical speculations have to strengthen our faith in immanent democratic potentials, especially when faced with intractable problems of senseless suffering, moral tragedy, and radical evil, that most readily reveal the ethical value of such speculative metaphysics. The speculative adventurousness of ecstatic naturalism most clearly reveals its pragmatic cash-value even if remaining a rather austere intellectual undertaking.

From the philosophical perspective of "ecstatic naturalism,"[9] responding to the "problem of evil" is not a problem of abstract theodicy, understood as a kind of "defense speech" for a transcendent, eternally unchanging, and ontologically perfect deity, but instead becomes a vital matter of posing an existentially embodied query aimed at creatively realizing sacred semiotic potentials within an ethically neutral at best, or far too often violently chaotic *chora* of nature naturing. What Corrington calls the "material maternal," or the always unsettling and decentering "unconscious" of nature naturing, resonates profoundly with the idea of an ontological ground of all ordinal natural complexes as *natura naturans*—what could be generally encapsulated in much of pre-Qin philosophical discourse with the Daoist expression "way-making emulates spontaneous emergence" (*dao fa ziran* 道法自然).[10]

Indeed, all is nature—on either side of the ontological divide. The so-called "ontological difference" as Corrington sees it, can be optimally understood as the endlessly spontaneous self-emergence of *natura naturans* as it is giving rise to various "natural complexes" and an infinitely diverse array of metaphysical "traits" within the pluralizing domains of *natura naturata* as emergent patterns of evolution. Corrington's metaphysical-theological project of "ecstatic naturalism" can be read as a kind of poetic phenomenology, speculative metaphysics, and embodied semiotics that offers an inexhaustible vocabulary and expansive imaginative horizon from which to approach basic questions as to how to more sustainably realize consummatory experiences or ecstatic events of sacred folds within a perpetually precarious cosmos.

A kind of non-foundationalist naturalistic and pragmatist cosmology permeates this ecstatically naturalist approach to theodicy. As already indicated, in addressing the problem of evil and senseless suffering within a theological perspective of "deep pantheism,"[11] one can feel free to completely pass over the traditional problematic of defending an omniscient, omnipotent, and omnibenevolent deity against charges of ontological inconsistency and ethical hypocrisy. A deep pantheist or ecstatic naturalist position focuses our ethical concern instead upon realizing a vitalizing description of embodied experience that can heighten our sensitivities to the ever-present potentially ecstatic and sacred dimensions of creatively democratic biosemiosis. At the same time, however, ecstatic naturalism offers a clear-eyed, yet edifying, means of facing up to the realities of ethically indefensible and theologically unforgivable aspects of that very same experience—the all-too-human forms of cruelty and tragedy at the core of far too many everyday natural transactions.

CONFUCIAN RELIGIOSITY AND ECSTATIC NATURALISM: KEEPING THE CONVERSATION GOING

In the remainder of this chapter, I will simply cast some exploratory light upon some of the ways in which an ecstatically naturalistic perspective of deep pantheism resonates with aspects of an early Confucian tradition of thinking about the emergence of ethical subjectivity against the backdrop of our embodied roles and relationships occurring within an anthropocosmic and correlatively co-creative cosmos. Confucian cosmology need not presuppose the rather parochial sounding notion of an "ontological difference"; at least insofar as this idea is understood in a post-Heideggerian fashion as a grand metaphysical narrative of discovery and decline of human *dasein* within a reified notion of a "Western" onto-theological destiny. However, I am suggesting that Corrington's potent philosophical theology of ecstatic naturalism gives us a very fecund discursive method—an oeuvre that is admirably open to constant evolution by remaining receptive to novelty, revision and transformation—by which to "approach" an interfaith conversation with a Confucian tradition of "theodicy." What we can achieve via a careful reading of Corrington alongside Confucian texts is the development of an interpretive methodology or an *ars contextualis* that avoids both the Scylla of "reductive naturalism"[12] and the Charybdis of supernatural or metaphysically bad "transcendent" theodicies.[13]

Indeed, much of the history of Western translations and interpretations of the Ru tradition have been a series of attempts to pour Confucian wine into Judeo-Christian wine skins—or, at the very least, to try and shoehorn Confucian discourse into a philosophical-theological tradition grounded in a post-Constantinian marriage between Greek substance metaphysics and a Pauline carnal hermeneutics. This would be a most unhappy marriage indeed and one that traumatically dichotomizes living subjectivity by cleaving off a "transcendent" or "supernatural" *spirit* juxtaposed to a mess of chaotic desires in an untrustworthy, because fundamentally depraved, *flesh*. By retrospectively projecting this host of problematic dualisms and supernatural metaphysical assumptions onto the Confucian tradition and by (mis) translating ideographs like *tian* 天 as "Heaven," *dao* 道 as "the Way," *ren* 仁as "Benevolence" and *yi* 義 as "Righteousness" without any hermeneutic contextualization, what we end up with in approaching Confucianism is an exotic specimen of a second-rate or knock-off version of Christianity. Ultimately we seem to be just left with an uncritical reception of a postcolonial Confucianism as a kind of strange hybrid "religious"[14] tradition that is internally inconsistent and probably in need of some serious messianic salvation.[15]

Corrington's liberating philosophical theology, on the other hand, offers ways of creatively and responsibly engaging in interfaith and intercultural conversations with a Confucian worldview. It will always remain a hermeneutical impossibility to leap outside of our familiar cultural contexts and concomitant philosophical "common sense"[16]—transcending, as it were, our embodied *parole* to immediately inhabit an unfamiliar Si-Meng Confucian *langue*. A viable interpretive option then is to make use of conceptual resonances and semiotic cross-fertilization to open up spaces of productive interfaith conversations.

Corrington's biosemiotic phenomenology and depth theology allow us to get to such a space of interpretive possibilities from which we can seek to let the Confucian conceptual clusters regarding "creativity" (*cheng* 誠), "anthropocosmic sacredness" (*tian* 天), and "uniquely embodied relational potential" (*du* 獨)[17] come into more resolute focus for us as we attempt to forge a forward-looking Confucian vision upon a global stage of interfaith conversations aimed at cultivating communities of care in an era increasingly marked by global-scale violence and destruction. The rapid diminishing of sacred values resultant from an erosion of so many natural, cultural and ritual ecosystems under the auspices of a idolatrous worship of a pernicious philosophical fiction of a discrete, autonomous, foundational "Self"—an atomized abstraction at the belly of the beast of nihilistic neoliberal ideology that promotes capitalist efficiency and technological "progress" under the banner of radically libertarian values and alienating social structures—is indeed a most disturbing and lamentable turn of events and represents a tragic foreclosure of so many ameliorating possibilities.

The opening lines of the *Zhongyong* invite us to ponder how the project of becoming consummately human within a role-based ethical vision and a ritual-focused religiosity is fundamentally a matter of resolutely cultivating one's "*tian* ordered" dispositions in order to realize ecstatic consummations of value within ever-expanding "circles"[18] of semiotic significance:

> What *tian* 天 commands (*ming* 命) is called natural tendencies (*xing* 性); drawing out these natural tendencies is called the proper way-making (*dao* 道); refining the process of way-making is called education (*jiao* 教). As for this proper way-making, we cannot quit it even for an instant. Were it even possible to quit it it would not be proper way-making. It is for this reason that exemplary persons (*junzi* 君子) are ever so concerned about what is not seen, and ever so anxious about what is not heard. There is nothing more present than what is imminent, and nothing more manifest than what is inchoate. Thus, exemplary persons are constantly concerned about their relationally entangled uniqueness (*shenqiduye* 慎其獨也).[19]

The moment at which joy and anger, grief and pleasure, have yet to arise is called a nascent equilibrium (*zhong* 中); once the affects have arisen, that they are all brought into proper focus (*zhong*) is called harmony (*he* 和). This notion of equilibrium and focus (*zhong*) is the great root of the [moral] world; harmony then is the advancing of the proper way (*dadao* 達道) in the world. When equilibrium and focus are sustained and harmony is fully realized in relational experience, the heavens and earth are recognized in their proper places and all things can find ways to flourish in the world.[20]

The "inchoate" or "incipient" stirrings of the affective events (*qing* 情) of heartminding provide us with an experience of cosmic creativity wherein the consummation of any "event" (*wu* 物) is always a matter of "creativity" (*cheng* 誠).[21] The "root" of the world's self-consummating capability is likewise present in our potential to inhabit a ritualized "as if" (*jishen ru shen zai* 祭神如神在) world of positive growth in value and significance even "transcending" our fallible and fragile somaesthetic conditions as finite, mortal creatures. The embodying (*ti* 體) of ritual spaces and temporal structures (*li* 禮) commands us to take a perspective of holistic aestheticism in appreciating our biosemiotic potentials to transform destructive and messy habit-energy into virtuosic performances (*de* 德) that "fully participate" in the creative potentials of *tian* 天 and earth (*di* 地) ceaselessly unfolding as unsummed totality of the myriad events and cosmic processes (*wanwu* 萬物).[22]

By carrying out a kind of embodied reflection as meditative epoche in the context of an ordinal phenomenology of the selving process we can be "working out from within"[23] *this* very selving process (*shendu* 慎獨) to discover potentials to inhabit ritually transformative practices, thereby realizing relational dynamics that ecstatically call us out of ourselves (*tianming* 天命)—a kind of *ek-stasis* that demands an ongoing cultivation and affective attunement of an ethical responsiveness grounded in empathic concern engendered by a robustly embodied moral imagination (*shu* 恕).[24]

It is this counter-factual, subjunctive "as if" trait of the dynamically realized natural complexes of ritual harmony (*lizhihe* 禮之以和為貴)[25] that provides a thoroughly naturalistic outlook and religio-aesthetic sense of moral purpose without appealing to any transcendent teleology, metaphysical realism, or substantialist ontotheology. By offering a sacred semiotic perspective from which to "make sense" of our suffering, tragic limitations, and encounters with both ordinary and radical evil without having to make appeal to a supernaturalist theodicy, Corrington's conception of "sacred folds"[26] discoverable within natural complexes and constituting the exceptionally dense semiotic spaces of human selving processes opens up onto the possibilities of realizing various symbolic forms of the "beloved community" in history (e.g., the sacred "body of Christ" or the "body of Ru rituals") and recognizing "god-ing" or *tian* 天-ing as a distinct natural complex having its own specific semiotic density within a realized order of *natura naturata*.

A Confucian conceptual cluster focused around the ethically anthropocosmic influences of *tian* 天 offers a way to think through semiotic perspectives that tarry between the sacred spaces opened up by ritual technologies creatively integrating (*cheng* 誠) the potentially divergent orders of *tian* and humanity.[27] In some respects, the two distinct "traits" of creativity described here resemble the ontological difference between *nature naturing* and *nature natured*. However, a Confucian perspective on a ritually constructive and role-based creativity and embodied ethical agency offers a thoroughly immanent justification of "evil"—simply understood as the reality of missed opportunities for interpersonal growth and the persistence of coercive hierarchies of power and violence. Confucian theodicy then also promises "infinite" spiritual growth in terms of contributing to ancestral significance and the sacred potential to occupy one's ritually stipulated roles as made a vocation by *tian*. Similar to Corrington's recognition of the ethically indifferent *chora* as nature's under-conscious, the Confucian tradition has no need to explain away the precarity of embodied human experience in a perpetually perishing cosmos. Nevertheless, we can be availing ourselves of ritual technologies to mitigate and transform the more pernicious and terrifying aspects of such a perspective regarding the ontological "ground" of things.

MENGZI ON A *TIAN*-CHISELED MORAL BODY

The *Mengzi* offers many passages that shed much light on a distinctive Confucian vision regarding the redemptive potentials of senseless suffering and radical evil. Here we will only consider two significant passages.[28] First of all, there is the idea that *tian* is recognized by Mengzi as being *anthropocosmic*. *Tian* quite literally is conceived as "seeing and hearing with the eyes and ears of the multitude" and "does not speak."[29] Just as the *Zhongyong* vision of co-creativity invokes anthropocosmic participation in realizing ecstatic, sacred and consummatory values in the world, the *Mengzi* offers a theory of political legitimacy grounded in an immanently divine trait of true kings—the so-called "heavenly mandate" (*tianming* 天命) that distinguishes hegemons from political leaders that warrant the "people's aspirations" (*minxin*民心). Mengzi's rejection of a transcendent "will of *Tian*" in favor of an ecstatically immanent anthropocosmic legitimacy is similar in spirit to Confucius' claim that only "persons can embellish way-making, and it isn't the case that 'the Way' can embellish persons."[30] Despite rejecting any transcendent or supernaturalist notions of *tian*, the Mengzi is still offering a kind of "ecstatically naturalist" vision that presupposes an awesome capacity *tian* has to test and strengthen the characters of exemplary persons through a kind of divine dispensation of redemptive suffering for those who are ecstatically called to participate in the continuous anthropomorphizing efforts of *tian*'s ceaseless creativity (*shengsheng bu xi* 生生不息).[31]

The passage that immediately comes to mind in reference to this dynamic of *tian* dispensed "redemptive"[32] suffering is *Mengzi* 6B15:

> Mengzi said, "When Shun lived in the mountains [he experienced extreme difficulty and poverty]. . . . Thus when *tian* intends to give someone a great ethical task, it first submits their heartminds and intentions to bitter suffering, their bodies to hard physical labor, starving their carnal frames, and emptying out their *shen* 身 bodies. This occurs so that in future action they cannot be led astray, and so that they can be empathic and bear up under the weight of their moral natures (*dongxin renxing* 動心忍性). [Such redemptive suffering] is also beneficial in terms of improving upon what they were previously incapable of doing. It is only by constantly evaluating one's moral shortcomings, that one is able to make improvements. Pondering difficulties in the heartmind, carrying out ethical deliberations, and then acting. [Sacred potential] is subtly expressed in the countenance and in the timbre of voices. Thus it becomes expressive."[33]

Whether experiencing sorrow or joy, suffering or exaltation, the *Mengzi* exhorts us to always make the most of our experiential ingredients by striving to realize the sacredness inherent in any situation as a natural complex with ecstatic potential:

> Mengzi said, "To fully exhaust the potentials of the heartmind is to realize (*zhi* 知) one's natural dispositions. To realize one's natural disposition is to realize *tian*. Preserving the heartmind, cultivating the dispositions, this is how to serve *tian*. Whether short-lived or long-lived it makes no difference. Refining the *shen* body in obedience [to *tian's* command] is how one establishes their life [as an ethical project] (*li ming* 立命)."[34]

I would like to give Mengzi the last word here in reiterating the vital significance of "embodied reflection" or "biosemiotic theology" involved in our courageous cultivation of joyous dispositions in the arduous processes of empathically engaging with relational virtuosity in the anthropomorphizing ritualization of the cosmos; especially as such ethically valiant activity is carried on in direct response to so much senseless suffering and radical evil:

> Mengzi said, "The myriad things are all completed here in me. There is no greater joy than to reflect upon my *shen* body and realize such creativity (*cheng* 誠) at work. There is nothing that will bring one closer in proximity to seeking 'relational virtuosity' (*ren* 仁) than strenuously engaging one's embodied moral imagination."[35]

CONCLUSION

It is my hope that this chapter can help in keeping the conversation going between ecstatic naturalism and Confucian role ethics both as distinctive methods and unique horizons of philosophical interpretation. A suppressed premise I suppose of this chapter is that the worst thing about traditional theodicies grounded in ontotheologies of Absolutism, linear, single-source thinking, and strict ontological transcendence is that they not only fail to provide accountability for senseless suffering, and violent acts of radical evil, but that they also tend to provide cover up for those who shoulder the most responsibility for the violence, oppression, and cruelty in the first place. Failing to ameliorate situations often results from a lack of moral imagination and a courage to act in creative democratic solidarity with and for others to ensure that a better world is possible. On these two fronts we certainly cannot fault the conversational horizons disclosed by Robert Corrington's project of ecstatic naturalism and the living tradition of Confucian role ethics. Perhaps though, similar to what the Stone Pass Gatekeeper said of Confucius, when it comes to cultivating moral courage and a radical political imagination we could hardly do much better than "knowing it to be impracticable, but still acting anyway."[36]

NOTES

1. Lydia Liu's *Translingual Practice: Literature, National Culture, and Translated Modernity—China, 1900–1937* (Stanford University Press, 1995) offers a powerful critique of unilateral and culturally homogenizing theories of translation, while more concerned with ways in which Chinese literary modernity can remain a heterogeneous and manifold play of the radical political imagination, Liu's critique of unilateral source target translation schema is reminiscent of A. N. Whitehead's unmasking of the "fallacy of the perfect dictionary" wherein every noun in a natural language is presumed to pick out one and only one natural object or a clear and distinct abstract concept. In place of such modular and mechanistic views of translation Liu offers 'translingual practice' as a dynamic and ethically conversational alternative that relies instead on the conceptual metaphors of "guest" and "host" languages: The idea of source language often relies on concepts of authenticity, origin, influence, and so on, and has the disadvantage of re-introducing the age-old problematic of translatability/untranslatability into the discussion. On the other hand, the notion of target language implies a teleological goal, a distance to be crossed in order to reach the plenitude of meaning; it thus misrepresents the ways in which the trope of equivalence is conceived in the host language, relegating its agency to secondary importance. (Liu, 27)

2. For more on metaxalogical thinking see William Desmond's *The Intimate Universal: The Hidden Porosity among Religion, Art, Philosophy and Politics* (Columbia University Press, 2016). Desmond sees the metaxological in global philosophy as going beyond either-or dialectics by maintaining a type of intimate presence in spaces of betweenness. For Desmond, any "univocal 'either-or' is to be dialectically dissolved as a fixation, in order to reveal the promise concealed in the

fixation" and this promise is always "transdialectical and points toward a metaxology of the intimate universal" (Desmond, 12).

3. Translations unless otherwise marked are my own. Cf. *Mengzi* 7A1: 孟子曰：「盡其心者，知其性也。知其性，則知天矣。存其心，養其性，所以事天也。殀壽不貳，修身以俟之，所以立命也。」Yang Bojun 楊伯峻 ed.《孟子譯注》(Beijing: Zhonghua Shuju, 1960), 301.

4. See Catherine Keller's *Face of the Deep: A Theology of Becoming* (Routledge, 2003) on tehomic plurisingularity that deeply challenges any puritanical origin stories and the domineering theologies of fixed teleologies that follow. Keller's work generally departs from any foundationalist mythologies in epistemology and ethical theory and offers a reading of the plurisingular fluidity of the name and accompanying grammatology of 'Elohim': "Such a theology, neither monistic nor dualistic, prepares a pluralism not of many separate ones but of plurisingularities, of interdependent individuations, constantly coming, flowing, *through* one another" (Keller, 179). 'Plurisingularity' as unique relational entanglement is precisely how I'm understanding ethical individuation (*shendu*慎獨) in the early Confucian tradition.

5. The phrase *tianrenheyi* is pervasive in the Confucian tradition and is usually rendered something like "heaven and human unification" and this can be traced to Tu Weiming's framework of interpretation. For Tu the "profound person" (*junzi*君子) has the chance to be an embodiment of a unique focal center for radiating webs of vital relationality as part of a "fiduciary community." Hence, the Zhongyong articulates a "metaphysics which advocates that the ultimate reality is perceivable and realizable in the moral life of every person because human natures is potentially a genuine manifestation of that reality" (Weiming杜維明 Tu, *Centrality and Commonality: An Essay on Confucian Religiousness* (State University of New York Press, 1998), 70). The use of "anthropocosmic" I appropriate from Tu and is in this chapter a way of saying how a humanized cosmos regulated by ritual propriety is achieved in a Confucian perspective. The use of ritual technologies quite literally *constructs* a harmonious social order while projecting "humaneness" (*ren* 仁) into the cosmos as a kind of *ecstatic naturalism*. For more on "ritual constructivism" see Michael Puett, "Ritual Disjunctions: Ghosts, Philosophy, and Anthropology" in *The Ground Between: Anthropologists Engage Philosophy*, eds. Veena Das, Michael Jackson, Arthur Kleinman, Bhrigupati Sing (Duke University Press, 2014). The basic idea is that it is a mistake to read early Confucian ritual theory as being about static teleology or "mirroring" an idealized cosmos. Rather, what ritual theory-praxis does in a Confucian worldview is perform iterations of subjunctive or disjunctive reversals as imaginative role playing that allows for transforming the "institutions" (*zhidu*制度) of ritual and music—which actually amounts to profound paradigm shifts in the historical inertia of received or sedimented structures of feeling in ethical and political culture.

6. The "Si-Meng lineage" (思孟學派) of Confucianism refers to a tradition associated with the purported author the the "Focusing the Familiar" (*Zhongyong* 中庸) text—Confucius' grandson, Zisizi (481–402 BCE)—and the *Mengzi*, a text traditionally thought to be compiled by Mengzi (372–289 BCE) and his disciples during the Warring States. Recent archaeological discoveries, like the Guodian Chu Bamboo Slips, have initiated a radical reappraisal of the Ru tradition and provide a

lot of "new" material for scholars to work with in continuing to articulate this living tradition.

7. See Roger T. Ames "What Ever Happened to 'Wisdom'? Confucian Philosophy of Process and 'Human Becomings'" in *Asia Major*, Third Series, Vol. 21, No. 1 *Essays in Honor of Nathan Sivin* (Academia Sinica 2008).

8. Richard Rorty's secular humanist faith in democracy and vision of a post-metaphysical culture committed in all areas to pragmatic anti-foundationalism can be found in his "Pragmatism as Romantic Polytheism" in *The Revival of Pragmatism: New Essays on Social Thought, Law, and Culture*, ed. Morris Dickstein, (Duke University Press, 1998), 21–36.

9. Robert Corrington considers "ecstatic naturalism" to be a "convergence and novel coalescence of pragmatism, naturalism, and postmodern psychoanalysis 'that' provides fertile soil for ordinal phenomenology in its continual exploration of the features of the world and nature." Robert Corrington, *Ecstatic Naturalism* (Indiana University Press, 1994), 15. He highlights the prospective pragmatic, and indeed prophetic, vision of a combination and transformation of pragmatism, naturalism, and postmodernism "as they converge within ecstatic naturalism [that] mark[s] a fundamental turning point within philosophy and theology. Of course, it is easy, and all too common, to claim that any given enterprise represents some kind of depth transformation of thought. The proper response to such claims is always that of the pragmaticist: by their fruits ye shall know them." Corrington, *Ecstatic Naturalism*, 15.

10. Corrington appeals to Baruch Spinoza's *Ethics* (read through a rather Nietzschean-Deleuzean colored lens) to make sense of the "ontological difference" between *natura naturata* and *natura naturans*. But he could've just as well turned to the *Daodejing* ch. 25: "There was some process that formed spontaneously/Emerging before the heavens and the earth. / Silent and empty, / Standing alone as all that is, it does not suffer alteration. / [All pervading, it does not pause.] / It can be thought of as the mother of the heavens and the earth. / I do not yet know its name (*ming*). / If I were to style it, / I would call it way-making (*dao*). / And if forced to give it a name, / I would call it grand. / Being grand, it is called passing, / Passing, it is called distancing. / Distancing, it is called returning. / Way-making is grand, The heavens (*tian*) are grand, / The earth is grand, / And the king is also grand. / Within our territories / There are four "grandees" / And the king occupies one of them. / Human beings emulate the earth, / The earth emulates the heavens, / The heavens emulate way-making, / And way-making emulates what is spontaneously so (*ziran*).

有物混成，先天地生。寂兮寥兮，獨立不改，周行而不殆，可以為天下母。吾不知其名,字之曰道,強為之名曰大。大曰逝,逝曰遠,遠曰反。故道大,天大,地大,王亦大。域中有四大,而王居其一焉。人法地,地法天,天法道,道法自然。(Ames and Hall trans., 2003:115).

11. For more on a deep pantheism see the chapters on "Selving" and "God-ing and Involution" in Robert Corrington, *Natures Sublime: An Essay in Aesthetic Naturalism* (Lexington, 2013).

12. For such a reductively naturalistic critique of classical Confucianism see Donald Munroe, "Mencius and an Ethics of the New Century," in *Mencius: Contexts and Interpretations*, ed. Alan K. L. Chan (University of Hawaii Press, 2002), 305–315. Also,

for an insightful response to Munroe's position see James Behuniak Jr. "Naturalizing Mencius" in *Philosophy East and West* 61, no. 3 (July 2011): 492–515.

13. For a *locus classicus* of such "Christianized" readings of Confucianism see James Legge's *The Chinese Classics with A Translation, Critical and Exegetical Notes, Prolegomena, and Copious Indexes* (Oxford, 1893). For more recent versions of this tendency to import Western assumptions into a "Confucian" packaging see a lot of the literature that purports to identify Confucian ethics as a type of Virtue Ethics complete with a metaphysical teleology. See, for example, Andrew Plak's *Ta Hsueh and Chung Yung: The Highest Order of Cultivation and On the Practice of the Mean* (Penguin Classics, 2004).

14. For some very interesting studies on the problematic construction of Confucianism as a world religion see Anna Sun's *Confucianism as a World Religion: Contested Histories and Contemporary Realities* (Princeton University Press, 2013) and Filippo Marsili's *Heaven is Empty: A Cross-Cultural Approach to "Religion" and Empire in Ancient China* (SUNY Press, 2018).

15. James Legge, as a good Presbyterian missionary, is quite the honest translator and offers a stinging indictment of the *Zhongyong* as a "blasphemy of the first order": "It begins sufficiently well, but the author has hardly enunciated his preliminary apothegms, when he conducts into an obscurity where we can hardly grope our way, and when we emerge from that, it is to be bewildered by his gorgeous but unsubstantial pictures of sagely perfection. He has eminently contributed to nourish the pride of his countrymen. He has exalted their sages above all that is called God or is worshipped, and taught the masses of the people with them they have need of nothing from without. In the meantime it is antagonistic to Christianity. By-and-by, when Christianity has prevailed in China, men will refer to it as a striking proof how their fathers by their wisdom knew neither God nor themselves." (Legge, 55)

16. See William James's "Common Sense" in *Pragmatism and Other Writings* (New York: Penguin Books, 1907/2000), and Friedrich Nietzsche on "the strange family resemblance of all Indian, Greek, and German philosophizing" in "On the Prejudices of Philosophers" in *Beyond Good and Evil* (1886).

17. Roger Ames and David Hall remind us that the "performance of *li* must be understood in light of the uniqueness of each participant and the profoundly creative project of becoming a person. *Li* involve an ongoing process of personal investment that, with persistence and effort, refines the quality of one's communal transactions. This process is dispositional and productive of identity-in-action" (Ames and Hall, 42). It is important that we recognize that what *li* 禮 and *du* 獨 give us is a kind of *insistent particularity* and *personal uniqueness* as relational achievement concepts, and not some sense of atomized individuality understood in a framework of universal human values.

18. The idea of realizing ecstatic experiences is similar to John Dewey's notion of "consummatory experience." See *Art as Experience* (Illinois University Press, 1934/1989). Dewey's aesthetic naturalism and phenomenology of experience grounded in a continuous rhythm of closure and disclosure is clearly indebted to Ralph Waldo Emerson's notion of "ethical perfectionism" as ceaseless self-transcendence occasionally marked by consummatory cadences. See his "Circles" in

Essays (1841) and also Naoko Saito's *The Gleam of Light: Moral Perfectionism and Education in Dewey and Emerson* (Fordham University Press, 2005).

19. Here the idea of "uniqueness" (*du* 獨), I wish to suggest, is best understood as "ritual potential" or the ability to inhabit or embody certain ritual practices and ritually stipulated roles that allow the subject to "transcend" a finite and messy "human nature" in order to inhabit a perpetually hopeful condition of continuous ethical-spiritual growth and immortal influence.

20. 天命之謂性，率性之謂道，修道之謂教。道也者，不可須臾離也；可離非道也。是故君子戒慎乎其所不睹，恐懼乎其所不聞。莫見乎隱，莫顯乎微，故君子慎其獨也。

喜怒哀樂之未發謂之中。發而皆中節謂之和。中也者，天下之大本也。和也者，天下之達道也。致中和，天地位焉。萬物育焉。Cf. Ames and Hall, 89–90.

21. See ZY 25: "Creativity (*cheng* 誠) is self-consummating (*zicheng* 自成), and its way (*dao* 道) is self-directing (*zidao* 自道). Creativity is a process (*wu* 物) taken from its beginning to its end, and without this creativity, there are no events. It is thus that, for exemplary persons (*junzi* 君子), it is creativity that is prized. But creativity is not simply the self-consummating of one's own person; it is what consummates events. Consummating oneself is authoritative conduct (*ren* 仁); consummating other events is wisdom (*zhi* 知). This is the excellence (*de* 德) of one's natural tendencies (*xing* 性) and is the way of integrating what is more internal and what is more external. Thus, whenever one applies this excellence, it is fitting." 誠者自成也。而道自道也。誠者物之終始；不誠無物。是故君子誠之為貴。誠者非自成己而已也，所以成物也。成己，仁也；成物，知也；性之德也，合外內之道也，故時措之宜也。Cf. Ames and Hall, 2001: 106.

22. See ZY 22 for "participating in the triadic transformations" 贊天地之化育， 則可以與天地參矣。Also ZY 26 for the idea of "ceaseless creativity and immortality" 故至誠無息，不息則久，久則微，微則悠遠，悠遠則博厚，博厚則高明。博厚所以載物。高明所以覆物也。悠久所以成物也。博厚配地，高明配天，悠久無疆. Cf. Ames and Hall, 2001: 105. For an excellent overview and deep philosophical hermeneutics of the various notions of embodiment on offer in early Confucian thinking see Deborah Sommer's "Boundaries of the *Ti* Body" *Asia Major* 21.1 (2008): 293–324.

23. For more on this idea of situated ethical cultivation as reorienting from within see Peter Hershock's *Buddhism in the Public Sphere: Reorienting Global Interdependence*: "Buddhist practice also does not proceed by invoking or valorizing particular positions transcendent to our world as presently organized and ordered—identifying, in essence, a goal or ideal toward which we are to work. Instead, Buddhist practice proceeds resolutely from the midst of our situation, precisely as it has come to be. It is not a process of *working toward* a given (effectively, predetermined) *destination*, but rather *working out* in a distinctively different *direction*. Buddhist practice—in any of its versions—begins in the immediate here and now and works at qualitatively transforming and reorienting the change dynamics and relational patterns playing out therein." (Hershock, 4) I would argue that the exact same thing can be said about Confucian self-fashioning or "sacred selving" in a ritualized anthropocosmic process of *tian*-based creativity.

24. See *Analects* 4.15 子曰:「參乎! 吾道一以貫之。」曾子曰:「唯。」子出。門人問曰:「何謂也?」曾子曰:「夫子之道, 忠恕而已矣。」And *Analects* 15.24 子貢問曰:「有一言而可以終身行之者乎?」子曰:「其恕乎! 己所不欲, 勿施於人。」Cf. Ames and Rosemont, 92 & 189.

25. See The *Book of Rites Liji* 《禮記》 41.12: "The Classicists [Confucians] are made expansive through learning without exhaustion, resolute in praxis without tiring, meditatively absorbed in dwelling without impropriety, penetrating above without falling into conundrums; Ritual practice values harmony, is beautified by resolute action and trust. It is a method for navigating anxiety, by exalting the worthy the multitudes are made compliant, rough edges can be fine-tuned to fit. The inclusive style [of bodily comportment] is just like this."

儒有博學而不窮, 篤行而不倦; 幽居而不淫, 上通而不困; 禮之以和為貴,忠信之美, 優游之法, 舉賢而容眾, 毀方而瓦合。其寬裕有如此者。 Cf. ctext.org: https://ctext.org/liji/ru-xing/zh#n71371 [accessed 6/21/21].

26. See Robert Corrington's introductory materials to the First Annual Congress on Ecstatic Naturalism, referenced from Wesley J. Wildman's "Robert Corrington's Ecstatic Naturalism and the Social Construction of Reality" in *A Philosophy of Sacred Nature: Prospects for Ecstatic Naturalism*, eds. Leon Niemoczynski and Nam T. Nguyen (Lexington 2015), 89–103. Corrington introduces the term "sacred fold" to describe a core trait of ecstatic naturalism: "An ecstatic naturalism is a perspective that seeks to move toward an aesthetic phenomenology of nature's 'sacred folds'—special centers of numinous meaning and power that may be found throughout nature. . . . From nature's sacred folds emerges a fierce self-othering, nature naturing, where 'it' moves ecstatically ejecting semiotically dense momenta. Nature naturing is the inexhaustible well of nature's atemporal creating under-conscious." (Corrington quoted in Wildman, 90)

27. See ZY 20: "Creativity (*cheng* 誠) is the way of *tian* (天之道); creating is the proper way of becoming human (人之道). Creativity is achieving equilibrium and focus (*zhong* 中) without coercion; it is succeeding without conscious reflection. Freely and easily traveling the center of the way—this is the sage (*shengren* 聖人). Creating is selecting what is efficacious (*shan* 善) and holding on to it firmly." 誠者，天之道也；誠之者，人之道也。誠者不勉而中，不思而得，從容中道，聖人也。誠之者，擇善而固執之者也。Cf. Ames and Hall, 99–104.

28. Mengzi 5A5 and 6B15. Like Confucius, Mengzi says that "*Tian* does not speak—it simply reveals itself through deeds and affairs" (5A5). Cf. Yang: 219. He also attributes the virtuosic potentials of *ren* 仁 (relational virtuosity), *yi* 義 (optimal appropriateness), *li* 禮 (ritual propriety), *zhi* 知 (wisdom), and *sheng* 聖 (sagehood) to *Tian* 天 (7B24). The rule of a moral king is also likened to the rule of *Tian* with the idea of "heavenly mandate" (*tian ming* 天命) in 5A4. Cf. Yang, 215.

29. Mengzi 5A5. In the context of explaining how the "heavenly mandate" of political legitimacy passed on from Yao to Shun Mengzi argues that it was the "multitudes" (*min* 民) who "accepted" (*shou* 受) this state of affairs. He cites the *Great Pledge* 《太誓》 in the Zhou *Documents* to claim that "*Tian* sees as my people see; *Tian* hears as my people hear." 『天視自我民視，天聽自我民聽』(Yang, 215).

30. *Analects* 15.29: 子曰:「人能弘道, 非道弘人。」Cf. Ames and Rosemont, 190.

31. This phrase and the idea that the "exemplary persons are self-strengthening without respite" (君子自強不息) comes from the *Book of Changes*周易 (cf. Wilhelm/Baynes, 6). These ideas are similar to the vitalizing description of inexhaustible energy available to the full participant in the triadic transformation adumbrated in the more ecstatic depictions of aesthetico-ethical creativity in the *Zhongyong*.

32. I am using the phrase "redemptive suffering" quite broadly here to refer to any kind of suffering undergone which has the inherent potential to be "redeemed" as part of future experiences indicative of spiritual growth or enhanced moral strength. I consider Confucian role ethics to be fundamentally a religion of "healthy mindedness" and as such it probably has little patience or need for the idea of a soteriology of "buying back" persons from an originally depraved sin nature.

33. 孟子曰:「舜發於畎畝之中，傅說舉於版築之間，膠鬲舉於魚鹽之中，管夷吾舉於士，孫叔敖舉於海，百里奚舉於市。故天將降大任於是人也，必先苦其心志，勞其筋骨，餓其體膚，空乏其身，行拂亂其所為，所以動心忍性，曾益其所不能。人恒過，然後能改；困於心，衡於慮，而後作；徵於色，發於聲，而後喻。入則無法家拂士，出則無敵國外患者，國恒亡。然後知生於憂患而死於安樂也。」(Yang, 298) It is interesting to compare the end of this passage with *Mengzi* 7A21 on the ineffable expression of relational virtuosity in the jade-like countenance and sound of the exemplary person. Such tropes are also common in the recently excavated Guodian corpus, particularly in the "Five Modes of Proper Conduct" 《五行篇》 (Cook, 465–85).

34. 孟子曰:「盡其心者, 知其性也。知其性, 則知天矣。存其心, 養其性, 所以事天也。殀壽不貳, 修身以俟之, 所以立命也。」(7A1, Yang, 301)

35. 孟子曰:「萬物皆備於我矣。反身而誠，樂莫大焉。強恕而行，求仁莫近焉。」(7A4, Yang, 302) This passage resonates with the *Analects* on the immediate realization of 'consummate conduct' (*ren*仁) as the best circulating relational dynamic. Confucius makes a claim about the immediate accessibility of *ren* as part of desiring subjectivity: "The Master said: Is *ren* far off? Whenever I desire *ren*, I find it is immediately present." 子曰:「仁遠乎哉？我欲仁，斯仁至矣。」cf. *Analects* 7.30; Ames and Rosemont, 117).

36. *Analects* 14.38: 子路宿於石門。晨門曰:「奚自?」子路曰:「自孔氏。」曰:「是知其不可而┌之者與？」cf. Ames and Rosemont, 181.

REFERENCES

Ames, Roger and David Hall eds. *Focusing the Familiar: A Translation and Philosophical Interpretation of the Zhongyong*. Honolulu, HI: University of Hawai'i Press, 2001.

———. *The Making this Life Significant: A Philosophical Translation of the Daodejing*. London: Ballantine, 2003.

Ames, Roger and Henry Rosemont, Jr. *The Analects of Confucius: A Philosophical Translation*. London: Ballantine, 1999.

———. *The Chinese Classic of Family Reverence: A Philosophical Translation of the Xiaojing*. Honolulu: University of Hawai'i Press, 2008.

———. "What Ever Happened to Wisdom? Confucian Philosophy of Process and Human Becomings" in *Asia Major*, Third Series, Vol. 21, No. 1 *Essays in Honor of Nathan Sivin*. Academia Sinica, 2008.

Behuniak, James Jr. "Naturalizing Mencius" in *Philosophy East and West*, Vol. 61, No. 3 (July 2011): 492–515.

Cook, Scott顧史考. *The Bamboo Texts of Guodian: A Study & Complete Translation*. Cornell East Asia Series, Ithaca, NY: Cornell University Press, 2012.

Corrington, S. Robert. *Ecstatic Naturalism: Signs of the World*. Bloomington, IN: Indiana University Press, 1994.

———. *Natures Sublime: An Essay in Aesthetic Naturalism*. New York: Lexington, 2013.

Chan, Alan K. L. ed., *Mencius: Contexts and Interpretations*. University of Hawaii Press, 2002.

Desmond, William. *The Intimate Universal: The Hidden Porosity among Religion, Art, Philosophy and Politics*. New York: Columbia University Press, 2016.

Dewey, John. *Art as Experience*. Carbondale: Illinois University Press, 1934/1989.

Hershock, Peter. *Buddhism in the Public Sphere: Reorienting Global Interdependence*. London: Routledge, 2006.

James, William. *Pragmatism and Other Writings*. New York: Penguin Books, 1907/2000.

Keller, Catherine. *Face of the Deep: A Theology of Becoming*. London: Routledge, 2003.

Legge, James. *The Chinese Classics with A Translation, Critical and Exegetical Notes, Prolegomena, and Copious Indexes*. Oxford, Oxford University Press, 1893.

Liu, Lydia. *Translingual Practice: Literature, National Culture, and Translated Modernity—China, 1900–1937*. Stanford University Press, 1995.

Munroe, Donald. "Mencius and an Ethics of the New Century," in *Mencius: Contexts and Interpretations*. ed. Alan K. L. Chan. University of Hawaii Press, 2002.

Niemoczynski, Leon and Nam T. Nguyen eds. *A Philosophy of Sacred Nature: Prospects for Ecstatic Naturalism*. New York: Lexington 2015.

Plak, Andrew. *Ta Hsueh and Chung Yung: The Highest Order of Cultivation and On the Practice of the Mean*. Penguin Classics, 2004.

Puett, Michael. "Ritual Disjunctions: Ghosts, Philosophy, and Anthropology" in *The Ground Between: Anthropologists Engage Philosophy*, eds. Veena Das, Michael Jackson, Arthur Kleinman, Bhrigupati Sing. Duke University Press, 2014.

Rorty, Richard. "Pragmatism as Romantic Polytheism" in *The Revival of Pragmatism: New Essays on Social Thought, Law, and Culture*. ed. Morris Dickstein. Duke University Press, 1998.

Saito, Naoko. *The Gleam of Light: Moral Perfectionism and Education in Dewey and Emerson*. New York: Fordham University Press, 2005.

Sommer, Deborah. "Boundaries of the *Ti* Body" *Asia Major* 21.1 (2008): 293–324.

Sun, Anna. *Confucianism as a World Religion: Contested Histories and Contemporary Realities*. Princeton, NJ: Princeton University Press, 2013.
Tu, Weiming杜維明. *Centrality and Commonality: An Essay on Confucian Religiousness*. State University of New York Press, 1998.
Yang, Bojun楊伯峻 ed. 《孟子譯注》. Beijing: Zhonghua Shuju, 1960.

Chapter 14

We Have Met the Enemy, and He Is Us

The Nature of Evil and the Evil of Nature in the Anthropocene

Robert W. King

INTRODUCTION

According to Jedediah Purdy, author of *After Nature: A Politics for the Anthropocene* (2015), the Anthropocene "finds its most radical expression in our acknowledgment that the familiar divide between people and the natural world is no longer useful or accurate."[1] This development problematizes the notion of a purely natural Evil or Suffering separated off from the human environment of modern, industrial society. This chapter aims to further explore the consequences of the Anthropocene concept for considerations of an Evil and Suffering inherent in Nature, with some attention to Ralph Waldo Emerson and William James, both Stoic optimists who wrestled with the existence of Evil and Suffering in their worlds. Both wrote passionately about our need to recognize and engage with the Evil and Suffering in Nature and human experience, a moral imperative that resonates with ecstatic naturalism, and, in the Anthropocene, still more imperative.

NATURE AND HUMANS

Emerson defines Nature as "essences unchanged by man; space, the air, the river, the leaf." He distinguishes this pure natural realm from our social realm, where Nature is impacted by humans, the realm where man has "applied to the mixture of his will with the same things, as in a house, a canal, a statue, a picture." The realms of Nature and Society are, for Emerson, readily distinct, with human influence on Nature insignificant: "But his operations taken together are so insignificant, a little chipping, baking, patching, and washing, that in an impression so grand as that of the world on the human mind, they do not vary the result."[2] Thus, inherent Evil or Suffering in Nature may well exist but as a purely natural product.

The concept of the Anthropocene disrupts this neat duality of Nature and Society as well as Emerson's innocent view of our "operations on nature" as "a little chipping, baking, patching." Coined in the 1980s (quarried or emergent in the Industrial Age), the Anthropocene is defined as our new geological era, "a new era of the earth's history, when humans are a force, maybe *the* force, shaping the planet" (emphasis in original).[3] The human hand and influences are visible in our climate, seasons, bioregions, deadly extremes of weather, manifest in the cracks in the Antarctic ice shelf and in the shrinking Arctic ice cap—no pure wilderness, virgin rainforest, or ocean escapes the impact of the human—seven billion of them and counting.

EMERSON'S "FATE": THE NECESSITY OF EVIL

Emerson is a strong influence on Robert Corrington (and on Nietzsche, especially in terms of his *homo natura* and naturalistic ethics reconciled to "the good of evil").[4] Emerson writes in his late essay "Success" (1870) that "Nature knows how to convert evil to good." There is Evil in Nature, and to the extent that we are products of Nature, Evil in humans—"good comes from evil" both in Nature and humans—parallels evident in ecstatic naturalism with resonance between our unconscious and the unconscious of Nature within nature naturing. Nietzsche picks up on this as well, with Emerson a stronger influence than many suspects,[5] and with their nascent understandings of the unconscious/subconscious.

In his later essay "Fate" (1860), Emerson opens with the normative question regarding the conduct of life, "How shall I live?"[6] I use the term "normative" as the answer concerns the Good, the True, and the Beautiful, concerns for both the Economy of the Self and the Art of Life. But unlike earlier essays such as "Circles" (1841), where he maps an expansive progress of spiritual growth outward, this older, wiser Emerson dwells more on the dark side of experience:

> We must see that the world is rough and surly, and will not mind drowning a man or a woman, but swallows your ship like a grain of dust. . . . The habit of snake and spider, the snap of the tiger and other leapers and bloody jumpers,

> the crackle of the bones of his prey in the coil of the anaconda—these are in the system, and our habits are like theirs.[7]

Emerson catalogues natural disasters and disease:

> At Lisbon an earthquake killed men like flies. At Naples three years ago ten thousand persons were crushed in a few minutes. The scurvy at sea . . . the cholera, the small-pox, have proved as mortal to some tribes as a frost to the crickets.[8]

As we are in Nature, these disasters and disruptions are a part of experience. "The forms of the shark, the *labrus*, the jaw of the sea-wolf paved with crushing teeth, the weapons of the grampus [killer whale], and other warriors hidden in the sea, are hints of ferocity in the interiors of nature."[9]

For Emerson, we too, as a part of Nature, have internalized "hints of ferocity." We are at risk, though, if we "lose some vigor of wild virtue," if due to the social conventions of civilization these "swarthy juices . . . lose astringency and acid."[10] (A parallel might be found in Thoreau's concept of the "Wild"—"In wildness is the preservation of the world.")

But while his essay dwells on our "limitation" and "irresistible dictation" of an often cruel and arbitrary fate—"Nature is no sentimentalist"[11]—the tone is ultimately affirmative. We enjoy "Power" and free will, with creativity, thought, and imagination affirmed as sources of redemption.[12] Limitation and "the ring of necessity" are the mothers of invention:

> For though Fate is immense, so is Power, which is the other factor in the dual world, immense. If Fate follows and limits Power, Power attends and antagonizes Fate. We must respect Fate as natural history, but there is more than natural history. . . . Man is not order of nature . . . but a stupendous antagonism, a dragging together of the poles of the universe and in "relation" to the other orders of Nature.[13]

Cultivated by civilization, our "hints of ferocity" sublimate into a "will to power," an "ambition of individualism" that manifests Emersonian virtues: "Independence, spirituality the will to stand alone."[14] The antagonism of an adversarial relationship between humans and Nature may be gracefully resolved, as in ecstatic naturalism, with attunement and discernment.

A moral dimension is evident here, this will to Power viewed as beneficial to humankind, as progressive, not a rugged but a democratic individualism, which is all the more necessary given the Anthropocene's challenges to humankind. William James shares with Emerson both an awareness of Evil and a consequent drive for its amelioration via discernment, engagement, and a progressive politics.

WILLIAM JAMES: THE REALITY OF EVIL

For James, there are those who are blind to Evil, by temperament living in the smile-button state of the once-born, the healthy-minded soul who chooses "perversely to shut his eyes to [Evil] and hush it up."[15] James, in *Varieties of Religious Experience*, privileges—and at times envies—the experiences and worldview of the twice-born, sick souls. "Systematic healthy-mindedness, conceiving good as the essential and universal aspect of being, deliberately excludes evil from its field of vision."[16] "To the man actively happy . . . evil simply cannot then and there be believed in. He must ignore it; and to the bystander he may then seem perversely to shut his eyes to it and hush it up," benignly to "let it grow into a deliberate religious policy."[17]

James, in his lecture on "The Religion of Healthy-Mindedness," initially seems sympathetic to "the deliberate adoption of an optimistic turn of mind": "What can be more base and unworthy than the pining, puling, mumping mood, no matter by what outward ills it may have been engendered? What is more injurious to others?"[18]

He distinguishes between two basic temperaments engaging with a sense of Evil: one party comprises "those for whom evil means only a mal-adjustment with *things*, a wrong correspondence on one's life with the environment" (emphasis in original). This sense of Evil is readily resolved or cured "merely by modifying either the self or the things, or both at once."[19] The other view, "more extreme . . . more instructive," considers Evil "more radical and general, a wrongness or vice in [our] essential nature, which no alteration of the environment, or any superficial rearrangement of the inner self, can cure, and which requires a supernatural remedy."[20] These adherents are James's "sick souls" and he directs his inquiry into "whether pity, pain, and fear, and the sentiment of human helplessness may not open a profounder view and put into our hands a more complicated key to the meaning of the situation," that is, the problem of Evil.[21]

> There is no doubt that healthy-mindedness is inadequate as a philosophical doctrine, because the evil facts which it refuses positively to account for are a genuine portion of reality; and they may after all be the best key to life's

> significance, and possibly the only openers of our eyes to the deepest levels of truth.[22]

For James, in the "normal process of life," in the game of life, "Radical evil gets its innings and takes its solid turn."[23] In language that echoes Emerson, he offers his own catalogue of Evil in Nature:

> Here on our very hearths and in our gardens the infernal cat plays with the panting mouse or holds the hot bird fluttering in her jaws. Crocodiles and rattlesnakes and pythons are at this moment vessels of life as real as we are; their loathsome existence fills every minute of every day that drags its length along; and whenever they or other wild beasts clutch their living prey, the deadly horror which an agitated melancholiac feels is the literally right reaction on the situation.[24]

Given the reality of Evil, a genuine philosophy must deal with it:

> Since the evil facts are as genuine parts of nature as are the good ones, the philosophic presumption should be that they have some rational significance and that systematic healthy-mindedness, failing as it does to accord to sorrow, pain, and death any positive and active attention whatever, is formally less complete than systems that try at least to include these elements in their scope.[25]

"The completest religions would therefore seem to be those in which the pessimistic elements are best developed."[26] He includes here Buddhism and Christianity, "essentially religions of deliverance: the man must die to an unreal life before he can be born into the real life."[27] We may include ecstatic naturalism in this pantheon for the twice-born.

For James (and Emerson and Corrington) there is a moral imperative here. Evil is a challenge, connecting us to the natural world, the Creation, and our communities as we endeavor to ameliorate Sufferings and tragedies. A Jamesian temperament recognizes Evil, pain, and loss, but seeks to improve the time, progress to a better world. In a letter to Henry James, he confesses, "I can't bring myself to blink the evil out of sight. . . . It's as real as the good."[28] But as his student and early biographer, Ralph Barton Perry, wrote, James was ready to "buckle on his armor and go forth to battle" with Evil. Redemption would come through transcendence, not from the world, but within Nature, a reconciliation within the mundane, immersion in experience and our physical existence.

EMERGENCE OF THE ANTHROPOCENE

In terms of buckling on our armor, the Anthropocene serves as a useful and inclusive concept that brings together an essential awareness of the various human interventions contributing to acidification and warming of the oceans, increased CO_2 in the atmosphere, extinction of species, thawing of the Arctic permafrost (releasing vast amounts of once-frozen and passive methane gas), negative consequences that especially impact lower economic classes.

It is both a political and a moral challenge, a large-scale matter of life and breath, life and death. Optimistically, the Anthropocene promises "a conceptual basis for environmental politics," identifying a problem so that we can now take responsibility for it and endeavor to comprehend it.[29] The concept creates an ethical impetus that melds well with the ethical purposiveness of Emerson, James, and of ecstatic naturalism.

Writing in the midst of the Industrial Age, the often prescient Friedrich Engels observed:

> Let us not . . . flatter ourselves overmuch on account of our human victories over nature. For each victory nature takes its revenge on us. . . . At every step we're reminded that we by no means rule over nature like a conqueror over a foreign people, but that we, with flesh, blood and brain, belong to nature, and exist in its midst, in that all our mastery of it consists in the fact that we have the advantage over all other creatures of being able to learn its laws and apply them correctly.[30]

What is the source of Evil and Suffering and to what extent are we humans now complicit? Language and definitions are necessary to the process of understanding cause, effect, and amelioration. How do we label, or find a general term, for the varieties of Sufferings—"Could it be 'Satan'?" For scientists, the Devil is largely our carbon footprint, and for some economists, these sins represent the necessary consequences of capitalism and satanic corporate billionaires manipulating our politics.

CONCLUSION

Thus, in the Anthropocene, we begin to understand the consequences of CO_2 and warming oceans and our complicity as our planet warms toward a Purgatory (or at least, everyone understands but those often holding political power in the United States). If global climate change is the "natural" result of global capitalism and its need for "cheap Nature"—cheap energy, raw materials, food—how do we manage the transvaluation of values, the necessary "greening" of American and global politics?

Nature is historical, but for capitalism the dualistic conception of Nature versus Culture enables the compartmentalization of Nature—it exists "outside" of Society and may therefore "be coded, quantified and rationalized to serve economic growth."[31] Capitalism as an economic system may be seen as "a way of organizing nature."[32] It is necessary to shift from the additive dualism of "humanity and nature" to the synthesizing compound "humanity-in-nature" and "nature-in-humanity."[33] To quote William James, "The trail of the human serpent is thus over everything."

In the progressive (and optimistic) vein of Emerson and James, Jedediah Purdy celebrates "the ideal of Anthropocene democracy," where an ecological self-consciousness will lead to a radical redistribution of political power, a sort of global Green Party United Nations: "Self-aware, collective engagement with the question of what kinds of landscapes, what kind of atmosphere and climate, and what kind of world-shaping habitation to pursue would all be parts of the repertoire of self-governance."[34]

There is always hope. One main point in terms of the Anthropocene is that, as human impact on the natural world grows more intrusive (placing "the scent of the serpent" on everything—we are the "Great Satan"), it should make us alert to possible shortcomings, when it comes to the problem of Evil and Suffering, of thinking exclusively in terms of a Nature-Human or Nature-Culture split, and to think in terms of "humanity-in-nature." In other words, and with respect to the Evils and Sufferings resulting from our global warming, "We have met the enemy, and he is us."

NOTES

1. Jedediah Purdy. *After Nature: A Politics for the Anthropocene* (Cambridge: Harvard University Press, 2015), 2–3.
2. Ralph Waldo Emerson. "Nature." Emerson, Ralph Waldo. *Nature and Selected Essays* (New York: Penguin Classics, 1983), 36.
3. Purdy, *After Nature*, 2.
4. See George Stack. J. "Emerson, Nietzsche, and Naturalistic Ethics." *Humanist* (November/December 1990): 21–25.
5. See ibid.
6. Emerson, "Nature," 361.
7. Ibid., 364.
8. Ibid., 364.
9. Ibid., 364–365.
10. George Stack. J. "Emerson, Nietzsche, and Naturalistic Ethics." *Humanist* (November/December 1990): 21–25, 46. 23.
11. Emerson, "Nature," 364.
12. Ibid., 375.

13. Ibid., 373.
14. Stack, 24.
15. William James. *The Varieties of Religious Experience: A Study in Human Nature*. Ed. Martin E Marty (New York: Penguin, 1985 [1902]), 88.
16. Ibid., 88.
17. Ibid., 88.
18. Ibid., 89.
19. Ibid., 134.
20. Ibid., 134.
21. Ibid., 136.
22. Ibid., 163.
23. Ibid., 163.
24. Ibid., 164.
25. Ibid., 165.
26. Ibid., 165.
27. Ibid., 165.
28. Phil Oliver. *Wiliam James's "Springs of Delight": The Return to Life* (Nashville: Vanderbilt University Press, 2001), 72.
29. Benjamin Kunkel. "The Capitalocene." *London Review of Books* 2 (March 2017): 22–28, 23.
30. Ibid., 24.
31. Purdy. *After Nature*, 24.
32. Ibid., 24.
33. Ibid., 24.
34. Ibid., 23.

REFERENCES

Emerson, Ralph Waldo. *Essays and Lectures*. Library of America, 1983.

———. "Nature." Emerson, Ralph Waldo. *Nature and Selected Essays*. New York: Penguin Classics, 1983. 35–82.

James, William. *The Varieties of Religious Experience: A Study in Human Nature*. Ed. Martin E Marty. New York: Penguin, 1985 [1902].

Kunkel, Benjamin. "The Capitalocene." *London Review of Books* 2 (March 2017): 22–28.

Oliver, Phil. *Wiliam James's "Springs of Delight": The Return to Life*. Nashville: Vanderbilt University Press, 2001.

Purdy, Jedediah. *After Nature: A Politics for the Anthropocene*. Cambridge: Harvard University Press, 2015.

Stack, George J. "Emerson, Nietzsche, and Naturalistic Ethics." *Humanist* (November/December 1990): 21–25, 46.

Index

About the Contributors

FOREWORD

Dr. Robert S. Corrington is the Henry Anson Buttz professor (emeritus) of Philosophical Theology in the Graduate Division of Religion at Drew University in Madison, New Jersey. He has developed an influential philosophical and religious theory of nature which he calls "ecstatic naturalism." Professor Corrington has spent the past several decades developing his own philosophical system of ecstatic naturalism, which continues to unfold in a series of books and articles. Early in his career he brought together the perspectives of Continental phenomenology and American pragmatism in order to benefit from the descriptive care given to lived experience by both traditions. Peirce and Heidegger were early interlocutors, later augmented with a deep appreciation for the work of Justus Buchler, whose own metaphysics represents a radical break with the various traditions of Western philosophy. In addition, he has long been in dialogue with South Asian philosophy, in particular *Advaita Vedanta* as expressed in the Upanishads. He has made several trips to India and is interested in theosophy.

EDITORS

Dr. Joseph E. Harroff received his PhD in comparative philosophy from the University of Hawai'i at Mānoa and an MA in Chinese Philosophy from the School of Philosophy and Social Development at Shandong University (山东大学). He teaches courses in philosophy, religion, and Asian studies at American University. His research engages with the creative interpretive horizons made possible for ethical and political theorizing via responsible intercultural contextualization, a world-encompassing hermeneutic phenomenology of relationally constituted persons, and philosophical translation as method. He has recently translated Zhao Tingyang's *All Under Heaven: The Tianxia System for a Possible World Order* (University of California Press, 2021).

Dr. Jea Sophia Oh is associate professor of Philosophy at West Chester University of Pennsylvania. She received a Master of Sacred Theology with a focus in feminist theology from Yale Divinity School and a Master of Philosophy and a PhD in Philosophical and Theological Studies from Drew University with a concentration in Women's and Gender Studies and Comparative Theology. Her research primarily focuses on the fields of Asian and comparative philosophies, religion and ecology, and postcolonial theory. Her first book *A Postcolonial Theology of Life: Planetarity East and West* (2011) is a path-making work in Korean ecofeminist theology and comparative philosophy. She is coeditor of *Nature's Transcendence and Immanence: A Comparative Interdisciplinary Ecstatic Naturalism* (2017). She is the chair and initiator of "Comparative Religion and Ecology" in the American Academy of Religion (MAR-AAR) and a board member of the "Women's Caucus" at AAR. She is the chair of the "Society for Asian and Comparative Philosophy" at the Central division of American Philosophical Association (APA). She also serves as the chair of the "Society of Study of Process Philosophy" at both Central and Eastern divisions of APA. She was recently appointed as the chair of the "International Society of Chinese Philosophy" at the Eastern division of APA.

CHAPTER CONTRIBUTORS

Katelynn E. Carver is a postgraduate researcher who studies process-relational metaphysics at the intersection of science, religion, literature, and culture. She has presented her research domestically and internationally, including presentations of original research for the American Academy of Religion and the Northeast Modern Language Association, and has spoken as a guest lecturer in Mülheim an der Ruhr, Germany, at the Second European Summer School for Process Philosophy and in the Department of Religion at Baldwin Wallace University in Berea, Ohio. In 2014, she earned her Master of Theological Studies focusing in both Religion and the Social Sciences and Religion, Literature, and Culture from Harvard University. She is currently a doctoral candidate at the University of St Andrews.

Desmond Coleman is a doctoral student in philosophical studies in religion at Drew University. The preliminary title of his dissertation is "Melanoses: Ruminations on Alchemy and Blackness," and will explore a variety of relations between the cultural history of alchemy, blackness (both racial and nonracial), German idealism, and contemporary black American popular culture and politics. He was a keynote speaker of the ninth International Ecstatic Naturalism Congress.

Susan Erck is a PhD candidate of Philosophy at the CUNY Graduate Center. She is a graduate teaching fellow in the Brooklyn College Department of Philosophy, as well as a writing across the curriculum fellow for Hunter College, and lecturer for Mississippi State University in the Department of Philosophy and Religion. She specializes in value theory, with particular interest in Economic, environmental, and women's issues. Her research involves the philosophy of causation in relation to methodology in social and political philosophy/theory. She is currently working on exploring the theme of ecological sustainability in Plato's *Republic* and other ancient Greek philosophical works. Her courses include "Ethics," "Social and Political Philosophy," "Business Ethics," and "Markets and Morality." She is a supporter of Prairie Protection Colorado and The Buffalo Field Campaign.

Dr. J. Edward Hackett is assistant professor of Philosophy at Southern University and A&M College in Baton Rouge, Louisiana, where he lives with his wife Ashley of fourteen years. He researches ethics, Continental philosophy, and American philosophy, which include attention to black philosophers and theologians who should be read as part of the canon of American philosophy. Recently, his work in American religious thought includes Emerson, Whitehead, Brightman, Benjamin Elijah Mays, and Martin Luther King Jr.'s roots in personalism. These explorations in American religious thought are also embedded, albeit unorthodoxly, in Hackett's 2019 novel, *Flight of the Ravenhawk* (2019), the first novel of a trilogy published by Ink Smith Press. Hackett is the author of *Persons and Values in Pragmatic Phenomenology: An Exploration of Moral Metaphysics* (2018) with Vernon Press and coedited *Phenomenology for the 21st Century* (2016) with Aaron Simmons published with Palgrave Macmillan and was the editor on *House of Cards and Philosophy* (2015). His writings on pragmatism and phenomenology have appeared in Spanish and Russian translation. In the 2020–2021 year, he has lectured to the Brooklyn Public library, India and Mexico. He will be speaking in the public author series at the East Baton Rouge Public Library in the next upcoming year. He received his PhD from Southern Illinois University in 2013 where he wrote a dissertation on the phenomenological ethics of Max Scheler, and he earned an MA in analytic philosophy from Simon Fraser University in 2008.

Dr. Robert W. King, Utah State University, has a background in American literature (especially Emerson and Thoreau) and American pragmatism, with an evolving interest in the nature and textures of modern religious experience. The literature of the desert and desert spirituality is a focus, the desert a fertile landscape for deep pantheism and ecstatic naturalism.

Marilynn Lawrence taught philosophy for a number of years in the greater Philadelphia area and currently works as a UX content strategist. She serves as a member of the Board of Directors for the International Society for Neoplatonic Studies where she has organized panels on topics such as providence and fate in Neoplatonism, the influence of other philosophical schools on Platonism in Late Antiquity. She has also presented a number of works related to ecstatic naturalism at the annual Ecstatic Naturalism Congress from 2013 to 2019, and is coeditor of *Nature's Transcendence and Immanence: A Comparative-Interdisciplinary Ecstatic Naturalism* (Lexington Books, 2017). Some of her other publications include "An Ecstatic Naturalist Approach to the Anthropocene," in *Deep Philosophy, Deep Ecology* (The Prometheus Trust, 2018); "Tension in the Soul: A Stoic/Platonic concept in Plutarch, Proclus, and Simplicius," in *Platonic Pathways* (The Prometheus Trust, 2018); "*Akrasia* and *Enkrateia* in Simplicius's Commentary on Epictetus's *Encheiridion*," in *The Neoplatonic Socrates* (University of Pennsylvania Press, 2014); "Who Thought the Stars Are Causes? An Exploration of the Astrological Doctrine Criticized by Plotinus," in *Metaphysical Patterns in Platonism* (The Prometheus Trust, 2014 reprint); "The Young Gods: The Stars and Planets in Platonic Treatment of Fate," in *Perspectives sur le néoplatonisme* (Les Presses de l'Université Laval, 2010); and "The Place of Chance or Fortune in Platonic Fate," in *Conversations Platonic and Neoplatonic* (Academia Verlag, 2010). Marilynn also practices painting in oil and other media and enjoys inline street and speed skating.

Dr. Eunkyoung Lee is assistant professor at Pai Chai University, South Korea. She received her BA in Christian Education and her MDiv in theology from Methodist Theological Seminary. She earned her doctorate degree in education science from Eberhard Karls Universität in Tübingen, Germany. Now, she has been doing her research with the topic of "humanistic education for the posthumans," persistently concerning on the role of religious education for newly emerging posthumans. Her recent publications include "An Exploration of the Meaning of Human Being in the Age of Posthuman Semiocapitalism: Being of Mutual Cooperation Rather than that of Competition" (2020), "Review of Mental Diseases as Social Diseases and an Exploration of a Possibility of Curing Them with Social Construction of Story-Telling: Concerning on 'The Community of Interpreters'" (2020), and "Education of Symbiosis(Co-living) for the Posthuman Age" (2020). She also translated Anton A. Bucher, *Psychologie der Spiritualität* (Weinheim: Beltz Verlag, 2007) as *Psychology of Spirituality* in Korean (2013).

Thomas Millary is a PhD student in the Rice University Department of Religion. He received an MA in theology from Drew Theological School where his research focused on Western esotericism and American philosophy, particularly on continuities between Hermeticism and the American philosophical tradition. He continues to study contemporary trends in American religion and Western esotericism, exploring how the intersection of espionage and occultism in the twentieth century helped give rise to the modern surveillance state.

Dr. Iljoon Park is a visiting professor of Christian Research Institute for Integral Studies at Methodist Theological University (MTU), Seoul, and a lecturer at Yonsei University. His research interests have focused upon the subjects of being-human in contemporary philosophies, cognitive science, evolutionary theories, evolutionary psychology, artificial life, and so on. His main publications include "Betweenness, the Illusory Self, and the Disruptive Subject: In Evolutionary Biology and Cognitive Science," *Tran-Humanities*, vol. 1 (2009), 177–199; "Betweenness and the Authentic Self: A Comparison of Daoist Thought with Heidegger within the Context of Living with an Authentic Selfhood," *Madang*, vol. 11 (2009), 95–117; "To Talk about Justice: Suturing the Whiteheadian Harmonious Subject to the Badiouan Militant Subject for the Universal Subject of ta me onta," *Korean Journal of Christian Studies*, no. 87 (2013), 303–319; "An Ontology of the between in Evolutionary Theories: The Ontology of Event in a Post-Human Age" in *Korean* (2014); "Evolutionary Theory and the Ontology of Event: Posthuman Implications of Evolutionary Theories" in *Korean* (2014); "Rereading of the Whiteheadian Understanding of Organism in a Transhuman Age: A Critical Review of the 'Extended Mind Theory,'" *Trans-Humanities*, vol. 8, no. 1 (2015), 111–130; and "A Post-naturalist Idea of Ec-stasy: an East-West Dialogue in a Trans-human Age," in *Nature's Transcendence and Immanence: A Comparative Interdisciplinary Ecstatic Naturalism*, eds. Marilynn Lawrence and Jea Sophia Oh (Lexington Book, 2018), 137–147.

Dr. Ji Young Park works across the fields of Sustainable Development, Biological Diversity, Transition Economies, Smart City, and values education. She is a research professor and instructor in the School of Economics at Yonsei University. She is currently a member of Sea of Gem Books and the position, senior policy chair. She received a BA in Economics from Sungkyunkwan University, as well as an MPA in Economics and a PhD in Economics from Yonsei University. She teaches introductory economics, macroeconomics, and microeconomics. Dr. Ji Young Park has a long-standing interest in biological diversity, dialogue and values, attitudes, Sustainable Economics. Before taking up the position of University Lecturer at University, she pursued an interdisciplinary research agenda working on projects in Economic, civil society, and Sustainable Development in ASEAN countries, SSA, Latin America, and South Korea. She is currently researching biological diversity and Smart city issues surrounding values and attitudes education and sustainable development.

Dr. Ick-Sang Shin is assistant professor of the Yeollim Academy of Liberal Arts at the Sungkonghoe University (SKHU) in Seoul, South Korea. He was born in South Korea in 1971. He received a bachelor's degree in Physics at the Seoul National University, and a master of divinity degree at the MTU in Seoul. He earned his PhD in Philosophy of Religion at the MTU and did postdoctoral work in the Institute for the Study of Theology at the SKHU. His primary research is focused on posthuman and climate crisis in the interdisciplinary field of Christianity, science, ecology, political philosophy, and the other religions.

Dr. Moon Son is currently teaching religious education at the Graduate School of Education at Yonsei University in Seoul, Korea. He received a BA, ThM, and PhD in Theology at Yonsei University. He has published many academic papers in the areas of public health (Covid-19), Christian higher education, religious education, sustainability, ecology, social justice, interdisciplinary perspective of religion and science, and convergence education of medicine and theology. He is an ordained pastor in the Korean Methodist Church and is serving Jesus Love Methodist Church as a senior pastor in Seoul Metropolitan City, Korea.

Dr. Jonathan Weidenbaum teaches courses in world religions, philosophy, and ethics at Berkeley College, New York City. Interacting with a largely international and culturally diverse student body, his experiences as an educator are a regular source of fuel for his intellectual passions. He has a PhD in philosophy (The University at Buffalo), and he specializes in topics pertaining to the philosophy of religion, classical American philosophy, and the prospects of a religious naturalism. More recently, he has done work in the philosophy of humor. Dr. Weidenbaum has published in *The Journal of Speculative Philosophy*, *Open Theology*, *Transcendent Philosophy*, *The Journal of Liberal Religion*, and many other journals and edited volumes. He has presented at conferences in Spain, Poland, India, Mexico, France, and all over the United States. Traveling is yet another inspiration for his philosophical work, and Dr. Weidenbaum's repeated journeys through Asia continue to make an impact on his writing.